W9-CLN-301

Experts who have actively participated in the drafting of this document:

Ross BARTLEY – Bureau of International Recycling (BIR)
Vittorio CIANI – European Commission
Eduardo MORERE MOLINERO – European Commission
Bernard DECKERS – Belgium
Dominique DELATTRE – International Atomic Energy Agency (IAEA)
Brian DODD – International Atomic Energy Agency (IAEA)
David HARVEY – United Kingdom
Nils HOLMBERG – Sweden
Atsushi ISHIKAWA - Japan
Michael ISAKOV – Russian Federation
Michael MATTIA – United States of America
Kiyotaka TERASHIMA - Japan
Gerard VAN DER REIJDEN – Netherlands

The work has been coordinated by

Bernard DECKERS – Belgium
and
Bernard ROUX-FOUILLET – United Nations Economic Commission for Europe
(UN/ECE)

United Nations Economic Commission for Europe	European Commission	International Atomic Energy Agency

Report on the Improvement of the Management of Radiation Protection Aspects in the Recycling of Metal Scrap

UNITED NATIONS
New York and Geneva, 2002

ECE/TRADE/278

UNITED NATIONS PUBLICATION

Sales No. E.01.II.E.22

ISBN 92-1-116789-2

CONTENTS

PREFACE 9

EXECUTIVE SUMMARY 11

   1.   General Presentation of the Issue 11

   2.   Recommendations for Improvement of the System 12

      A.   Introduction of discrete radioactive sources 12
      B.   Introduction of uncontrolled radioactively contaminated material 14
      C.   Introduction of material with a very low level of radioactivity, released in accordance with a national regulatory framework 15
      D.   Controls on the output materials from the metal works – metal, slag and off-gas dust 16

CHAPTER I. INTRODUCTION 17

CHAPTER II. OVERVIEW OF THE PRESENT SITUATION 21

   2.1. The Metal Recycling Loop 21

   2.2. Types of Metal Scrap 21

      2.2.1. Home scrap (own arising or revert scrap) 21
      2.2.2. New scrap (traded scrap) 22
      2.2.3. Old scrap 22

   2.3. Basic Information on Radioactivity 22

      2.3.1. Radioactivity and radiation 22
      2.3.2. Background radiation 23
      2.3.3. Origin of the radioactivity 23
      2.3.4. Discrete sources 23
      2.3.5. Radioactive contamination 24

   2.4. The Potential Introduction of Discrete Sources in Metal Scrap 24

      2.4.1. Primary and secondary metals production 24
      2.4.2. Designers, owners, operators and maintainers of plant and equipment 25
      2.4.3. Demolition of plant and discarding of equipment 25
      2.4.4. Scrap collectors and processors 26

   2.5. The Potential for Contamination of Scrap by Radioactively Contaminated Material 26

      2.5.1. Primary and secondary metals production 26
      2.5.2. Designers, owners, operators and maintainers of plant and equipment 27
      2.5.3. Demolition of plant and dismantling of equipment 27
      2.5.4. Scrap collectors and processors 27

   2.6. Present Situation Regarding the Limitations of Radioactive Substances in Scrap 27

      2.6.1. Regulatory requirements 27
      2.6.2. Contractual specifications applied in the industry - Business acceptance limits 28

   2.7. Responsibility in Case of Discovery of an Orphan Source or Radioactively Contaminated Material in Metal Scrap 28

2.8. Paucity of Information 29

CHAPTER III. PREVENTION OF CONTAMINATION OF SCRAP THROUGH REGULATORY CONTROL 31

3.1. General 31

3.2. Prevention of Occurrence of Orphan Sources 32

3.3. Prevention of Radioactive Contamination 33

3.4. Materials with Naturally Occurring Radioactive Materials 34

CHAPTER IV. DETECTION AND MONITORING OF RADIOACTIVE MATERIALS IN THE SCRAP WITHIN THE RECYCLING INDUSTRY 37

4.1. General Approach 37

4.2. Factors Affecting Detection 38

4.2.1. Radiation penetration 38
4.2.2. Background radiation 38
4.2.3. Methods of radiation detection 39

4.3. General Considerations when Monitoring Metal Scrap for Radioactivity 39

4.4. Monitoring Points 40

4.5. Fixed Monitoring Systems 40

4.5.1. General features 40
4.5.2. Detection of the presence of a vehicle and speed measurement 41
4.5.3. Detector heads 41
4.5.4. Computing hardware/software 43
4.5.5. Alarm criteria and false alarms 43
4.5.6. Peripheral equipment 44
4.5.7. Practical testing of system performance 44
4.5.8. Reliability 46

4.6. Portable Detector Systems 46

4.7. Visual Observation of the Scrap 47

CHAPTER V. ACTIONS TO RESPOND TO A DETECTION OF RADIOACTIVE MATERIALS IN THE SCRAP WITHIN AND OUTSIDE THE RECYCLING INDUSTRY 49

5.1. Responsibilities of the Parties 49

5.1.1. General description of scrap flow and economic operators in scrap trading 49
5.1.2. Determination of the ownership of scrap during trading and transport 49
5.1.3. Responsibility of operators of detection equipment 50

5.2. Actions to be taken when an Alarm Level has been Exceeded 51

5.2.1. Immediate actions to be taken when an alarm level has been exceeded 51
5.2.2. Determination of risk to human health or the environment 52
5.2.3. Subsequent actions to be taken 52

5.3. Containment of the Radioactively Contaminated Materials 53

5.3.1. Dispersible and non-dispersible radioactive substances.. *53*
5.3.2. Protection packages (short-term and long-term)... *53*
5.3.3. Transport packages *53*

5.4. Personnel Protection **54**

5.5. Decontamination **54**

5.5.1. Decontamination at the site of detection *54*
5.5.2. Decontamination in a facility *54*
5.5.3. Decontamination of soils and loose wastes *55*
5.5.4. Decontamination of personnel *55*

**5.6. Transportation (Material Transported After Detection, Characterization,
and Containment of the Radioactive Substances)** **54**

5.6.1. Conditions for the return of radioactively contaminated recyclable metal to the point of origin *55*
5.6.2. Transfrontier shipment *55*

5.7. Reporting **55**

5.7.1. Agencies to receive reports... *56*
5.7.2. Reporting persons.. *56*
5.7.3. Elements *56*
5.7.4. Reporting format *57*

CHAPTER VI. INTERNATIONAL COOPERATION ON ORPHAN SOURCE ISSUES......................**59**

6.1. The Need for Cooperation and Information Exchange..**59**

6.2. IAEA Action Plan Related to Events Involving Orphan Sources **59**

6.2.1. Missing and found source database... *60*
6.2.2. The IAEA radiation events database (RADEV) *60*
6.2.3. International catalogue of sealed radioactive sources and devices, including *61*
transport containers *61*
6.2.4. Other relevant information collection *61*

CHAPTER VII. FINAL CONTROLS ON THE PRODUCED METAL, SLAG AND OFF-GAS DUST **63**

7.1. The Aim of Radioactivity Control on Products of the Process...**63**

7.2. Detection Systems for Radioactive Contamination of Products **66**

7.2.1. Metal.. *66*
7.2.2. Slag *66*
7.2.3. Off gas dust.. *67*

**CHAPTER VIII. CONCLUSIONS AND RECOMMENDATIONS FOR THE IMPROVEMENT
OF THE SYSTEM** **69**

8.1. Conclusions...**69**

8.2. Recommendations for Improvement of the System **70**

8.2.1. Introduction of discrete radioactive sources.. *70*
8.2.2. Introduction of uncontrolled radioactively contaminated material................................. *71*
*8.2.3. Introduction of material with a very low level of radioactivity, released in accordance with a
national regulatory framework*.. *73*
8.2.4. Controls on the output materials from the metal works – metal, slag and off-gas dust *73*

ANNEX 1. TERMINOLOGY 75

ANNEX 2. IAEA RADIATION EVENTS DATABASE (RADEV).. 77

ANNEX 3. DESCRIPTION OF WHAT RADIOACTIVE SOURCES MIGHT LOOK LIKE 79

ANNEX 4. IAEA REGULATIONS FOR THE SAFE TRANSPORT OF RADIOACTIVE
          MATERIAL.. 85

ANNEX 5. SPANISH PROTOCOL FOR COLLABORATION ON THE RADIATION
          MONITORING OF METALLIC MATERIALS 91

ANNEX 6. IAEA CODE OF CONDUCT ON THE SAFETY AND SECURITY OF
          RADIOACTIVE SOURCES, SEPTEMBER 2000 101

PREFACE

Metals have become some of the most recycled materials in the world. The risk of recycling radioactive contaminated materials has greatly increased.

Especially during the past several years, the potential problems of radioactive contamination of commercial metals has become more visible and of more concern to both radiation professionals and those who might come in contact with the contaminated material. This concern has led to several programmes, both at the national and the international level. At the national level, some countries have installed radiation detection equipment at their borders in an attempt to prevent contaminated metal from entering the country. Others have developed more extensive protocols to deal with a wider variety of activities. The Spanish Protocol for Collaboration on the Radiation Monitoring of Metallic Materials (Annex 5 of this report) is one such protocol. In the United States, the Conference of Radiation Control Program Directors (CRCPD) has been working with various regulatory agencies to develop a programme to collect orphaned radiation sources, as well as develop materials and training programmes to help make the public aware of such sources and their potential dangers.

At the international level, there are several agencies attacking the problem. The International Atomic Energy Agency (IAEA) has a number of initiatives under way as part of its Action Plan for the Safety and Security of Radiation Sources. These are described in other parts of this publication. The United Nations Economic Commission for Europe (UN/ECE) has approached the problem from a slightly different perspective.

In order to develop and maintain a partnership between government authorities, the steel industry, the metal scrap recycling industry and competent authorities in the field of nuclear safety and radiation protection, the UN/ECE decided to organise in May 1999 a *Seminar on Radioactive Contaminated Metal Scrap* at the invitation of the Government of the Czech Republic.

Participants in the Seminar recommended the following:
> ➢ in order to facilitate the exchange of information, a register of missing radioactive sources should be established;
> ➢ cooperation should be improved between regulators and industries, at the national as well as the international level;
> ➢ amendments should be developed to existing regulations so that they become more appropriate for orphan sources and uncontrolled contaminated metal scrap;
> ➢ efforts should be pursued to develop/derive internationally accepted clearance levels;
> ➢ the issue of public perception/acceptance should be more seriously addressed ;
> ➢ responsibilities should be clearly defined in order to facilitate the application of the « polluter pays » principle ;
> ➢ the additional burden/cost of detecting and disposing of radioactive sources should be borne by the industry and by society in general, according to their responsibility to properly manage detected radioactive sources.

The Seminar further recommended that a Team of Specialists on Radioactive Contaminated Metal Scrap be set up under the auspices of the UN/ECE to serve as a forum where technically qualified partners of the steel industry, government representatives and qualified international organisations will consult and propose solutions acceptable at the international level in order to harmonise legislation, the systems of measurement, the levels of investigation concerning radioactivity content of metal scrap and, possibly draw up codes of practice/conduct in this area.

This Team of Specialists has produced the present «Report on the Improvement of the Management of Radiation Protection Aspects in the Recycling of Metal Scrap». This publication is aimed at persons who may not have much awareness of radiation science, but who may on a very infrequent basis come into contact with a stray radiation source or radioactively contaminated material during the normal course of business. It is most important for these persons to be able to recognise and understand the potential problem, to be able to take basic safety precautions, and know how to get professional assistance. It is strongly recommended that a source of professional assistance be located before a need arises, so that staff can quickly get assistance when radioactive material is encountered.

It is also aimed at the national authorities in charge of radiation protection and of the safety and security of radioactive substances. The publication contains a number of recommendations in order to improve the situation regarding possible contamination of scrap, part of them being addressed to these authorities.

This publication provides an overview of the scrap metal industry, how radioactive substances can become incorporated into scrap and finished metal, how to detect them, and how to respond when they are detected. A number of references are mentioned for the reader who may wish to study the issue further. It provides information on some national and international standards, requirements, and procedures that are valuable in determining the degree of problem that has been encountered. It also presents recommendations for prevention, detection, and reaction to radioactive material in scrap metal. In addition, it provides introductory information on monitoring and detection of radioactive materials in metals.

EXECUTIVE SUMMARY

1. GENERAL PRESENTATION OF THE ISSUE

In order to conserve natural resources, recycling of materials is becoming more and more important. Recycling actively contributes to sustainable development.

This is particularly the case in the metal industry, where recycling has a long history and has been carried out for over a century before sustainable development became an important issue. Compared with other materials, metals have the major advantage of not just being recyclable but of being infinitely recyclable. This advantage has been well recognised since the industrial revolution and currently about half of the steel production world-wide is made from metallurgical scrap. This percentage is still increasing, as it is for most other metals.

Although stringent controls are applied to radioactive substances, when there is a loss or lack of regulatory control, radioactive materials can be accidentally introduced into metallurgical scrap and hence into the metal production process.

International organisations have issued safety standards[1] and recommendations to keep radioactive sources or radioactively contaminated materials under strict control. Many countries have adopted these safety standards and recommendations in their national regulations. Nevertheless there are still situations where regulatory control is not effective or sources over which regulatory control has been lost. This is particularly the case in countries where the regulatory and technical infrastructure is not sufficiently developed. Since the metallurgical scrap market is a world-wide market, import of scrap may then cause some additional risk of introduction of radioactive materials into the recycled material.

In addition to regulatory control, the metal recycling and producing industries have organised themselves to reduce the probability of sources that escape regulatory control being introduced into the recycling loop. These measures are aimed at detecting such radioactive substances as early as possible, but their detection is not an easy task. Even with the most sensitive and sophisticated equipment, undetected radioactive sources or materials may be introduced into the recycling process. Systematic controls on measuring the radioactivity of products leaving the metal works have therefore proven necessary in order to guarantee their quality.

This report presents recommendations that should help avoid the introduction of discrete sources and improperly released radioactively contaminated material into the recycling stream. Taking into account such recommendations the probability of any undesirable introduction of radioactive materials into the recycling loop will be minimised. This will ensure better protection of the workers and the public.

The report also expresses the desire of the scrap recycling and metal production industry to be able to make informed decisions on the purchase and use of the material that is properly released from specific activities such as the nuclear industry.. This is to assure the customer of the recycled metal product that it meets desired specifications regarding radioactivity, and thereby maintain consumer confidence.

The report also reflects the concern of the industry as regards the costs arising from the detection or processing of radioactive substances that enter the recycling loop.

---

[1] "International Basic Safety Standards for protection against ionising radiation and for the safety of radiation sources", published by the International Atomic Energy Agency under the Safety Series n° 115. This document is jointly sponsored by the Food and Agriculture Organisation of the United Nations, the International Atomic Energy Agency, the International Labour Organisation, the Nuclear Energy Agency of the Organisation for Economic Co-operation and Development, the Pan American Health Organisation and the World Health Organisation.

2.   RECOMMENDATIONS FOR IMPROVEMENT OF THE SYSTEM

Radioactive material may be introduced in metallurgical scrap through three different pathways.

> Discrete radioactive sources may be introduced into the scrap, due to the fact that such radioactive sources may escape from regulatory control because they are abandoned, lost, misplaced, stolen or otherwise transferred without proper authorisation.

> Uncontrolled radioactively contaminated material may appear in the scrap stream from the process where the material has been used. The material may have become contaminated after contact with either natural radionuclides or man-made radionuclides. One example of this might be in an extraction industry where scale containing naturally occurring radioactive material (NORM) that is deposited in pipes or equipment may not be under regulatory control in the first place. Another example could be improperly released material that has been used in the nuclear industry and was contaminated with man-made radionuclides above regulated limits.

> Introduction of material with a very low level of radioactivity, released in accordance with - the national regulatory framework.

Three main topical areas related to the risk of introduction of radioactive materials into the scrap recycling process have been identified where improvements -should be made by international organisations, national Governments and industry, namely:

> Prevention of the introduction of radioactive materials.

> Detection of such an introduction (measurement, procedures and voluntary provision of information).

> Reaction capability to cope in the event of a detection of such an introduction.

These topical areas are developed hereafter for each of the three introduction pathways mentioned before.

A.   Introduction of discrete radioactive sources

Discrete radioactive sources may be introduced into the scrap, due to the fact that such radioactive sources may escape from regulatory control because they are abandoned, lost, misplaced, stolen or otherwise transferred without proper authorisation. Such sources could cause both external or internal exposure as well as significant economic impacts. External exposure can occur due to physical contact or close proximity to the radioactive material. Internal exposure can occur due to direct contact with or processing of any uncontained material that can cause a worker to inhale, ingest or absorb the radioactive material. In addition, the introduction of such material may generate significant business disruption and financial loss due to detection and rejection of the material or from the handling and processing of undetected material which could contaminate equipment, grounds, products and by-products.

Prevention

The main measure to prevent entry of radioactive material into metal scrap is to ensure adequate control over the discrete sources by users and the national regulatory authorities.

The IAEA has developed standards of safety for protection against ionising radiation and radioactive sources. Their application significantly contributes to the prevention of radioactive sources from becoming orphaned. All Governments are strongly encouraged to expeditiously implement or strengthen their own regulations based on these standards, and to strictly enforce these regulations. In particular, the Code of Conduct for the Safety and Security of Radioactive Sources is recommended for implementation by all States.

Detection

Scrap recycling and metal producing companies, with the assistance of trade associations, national radiological protection organisations and suppliers of radiation sources, are encouraged to:

- train their personnel;
- develop procedures for visual inspection of scrap, principally during collection, in order to find discrete sources at their point of entry to, or early in the recycling industry; and,
- install and use detection equipment according to the manufacturer's specifications and to the recommendations of this document. Detection equipment should be installed at crucial points in the recycling loop, in particular, prior to locations where handling, processing or melting of the scrap could damage a source.

Reaction

Sealed sources may well have labels and markings indicating past ownership, or certification, as well as serial numbers that give the possibility of tracing an original user. However, it is in the nature of the demolition sector that even if an investigation were undertaken, the original owner may turn out to be a now-defunct business.

The issue of the management and disposal of orphaned sources that have been discovered needs to be addressed urgently by each State's regulatory body. The IAEA Code of Conduct makes the statement that "Every State should ensure that its regulatory body ...is prepared, or has established provisions to recover orphan sources and to deal with radiological emergencies and has established appropriate response plans and measures". Some considerations in this respect include:

- guidelines for identifying and characterising such sources;
- arrangements for identifying appropriate destinations for managing their storage or disposal;
- adequate arrangements for transporting them to such destinations, including return of materials across national borders.

Competent authorities should make provisions to clearly allocate responsibilities associated with managing discovered orphaned sources. It is regarded as unfair to place the cost burden of storage and disposal of orphan sources, or clean-up of contamination caused by them, on the facility that finds them.

For new sources, the producer responsibility concept may also be used with the costs of the end-of-life management of sources internalised in their initial selling price. This removes the financial burden from the last owner or holder of source.

Several countries are providing a free-of-charge method of disposal for orphan sources as a means to encourage their detection and appropriate disposition. It is recommended that such a procedure be applied world-wide. In addition, efforts to require the return of sources to the supplier at the end of life are to be encouraged since this also decreases the probability of a source becoming out of regulatory control. The use of voluntary agreements and codes of practice as exemplified by the Spanish protocol (cf. annex 5) is recommended on a national and international basis

The scrap recycling and metal producing industry is encouraged to actively provide information to the reporting mechanisms set up by the IAEA following the discovery of orphaned sources. These data can then be analysed with a view to making further improvements and recommendations.

B.   Introduction of uncontrolled radioactively contaminated material

The introduction of uncontrolled material into the recycling stream that is radioactively contaminated with either natural radionuclides or man-made radionuclides could pose similar health and economic impacts as those for discrete sources. This would also cause significant business disruption and financial loss due to detection and rejection of the material or handling and processing of material which could contaminate equipment, grounds, products and by-products. However, such impacts would usually be of a lesser magnitude than those encountered with discrete sources but at a greater frequency.

Prevention

The main measure to prevent entry of uncontrolled radioactively contaminated material into the metal scrap stream is to ensure adequate control over the materials coming from these specific activities or industries. This has to be supervised by the national regulatory authorities based on the recommendations of the IAEA, as set up in its standards and guidelines, especially the Basic Safety Standards.

A special case concerns industries whose activities deposit NORM at detectable levels onto materials that could be recycled. This can be the case for example in the extractive industries, notably those dealing with oil, gas, coal and phosphate. Since NORM is generally not regulated, NORM contaminated materials are often freely sold into the open market. This undesirable situation is being addressed by the IAEA, which is rapidly developing further recommendations regarding which materials should come under the scope of regulatory control. In the meantime, it is a practical desire of the scrap metal industry that they be informed by those selling or disposing of such materials whenever there is good reason to believe that due to origin or function, the particular materials are likely to be NORM contaminated. This avoids the problems associated with rejecting material after NORM contamination has been detected at a receiving operation, perhaps after passing through several hands. It is desirable that such contaminated materials be specifically identified and kept separate from the normal scrap recycling circuit so it does not enter unrestricted metal products.

Detection

Scrap recycling and metal producing companies, with the assistance of trade associations, national radiological protection organisations are encouraged to:

- train their personnel;
- develop adequate procedures, principally during collection, for determination of possible NORM contaminated materials based on their origin or function;
- install and use detection equipment according to the manufacturer's specifications and to the recommendations of this document. Detection equipment should be installed at crucial points of the recycling loop, in particular prior to locations where handling, processing or melting of the scrap may present an exposure potential to workers or the potential for contamination of equipment, grounds, products or by-products.

Reaction

The metal recycling industry is seriously disadvantaged with regard to these materials and requires assistance. Its operations are unfairly bearing a major share of the costs of detecting, characterising, segregating, storing and disposing of contaminated materials. It would seem appropriate that the concept of the "polluter pays" principle also be used for management of uncontrolled radioactively contaminated material and radioactive contamination caused by these materials. In this respect, issues related to the proper disposition of materials discovered to be contaminated needs to be treated in a similar manner to that of discrete sources discussed above.

Some considerations in this respect include:

- guidelines for identifying and characterising such material;
- arrangements for identifying appropriate destinations for managing their storage or disposal;
- adequate arrangements for transporting them to such destinations, including return of materials across national borders.

Competent authorities should make provisions to clearly allocate responsibilities associated with managing the contaminated material. It is regarded as unfair to place the cost burden of storage and disposal of these contaminated materials, or clean-up of contamination caused by them, on the facility that finds them.

The concept of the 'polluter pays' principle may also be used for the management of such contaminated material, alternatively a free-of-charge method of disposal for such contaminated material is recommended. The use of voluntary agreements and codes of practice as exemplified by the Spanish protocol (cf. annex 5) is recommended on a national and international basis.

C.   Introduction of material with a very low level of radioactivity, released in accordance with a national regulatory framework

The introduction of low level radioactive material into the general recycling circuit, which is properly released according to the appropriate regulatory framework, is also of concern to the scrap recycling and to the metal producing industry. Because naturally occurring radioactive materials can cause alarm in the detection systems which might be considered from artificial origin, it is important to be able to trace the original supplier of the material to determine the appropriate response.

Prevention

Radioactive materials that are exempt from the requirements of the IAEA Basic Safety Standards or materials that are released from regulatory control do not have any significant radiological hazards associated with them. However, there is a perception that all radioactivity or all radiation is hazardous regardless of the level.

Therefore, as part of the contractual provisions and in order to satisfy the general customer demand, the metal recovery and recycling industry requires from the facility selling or disposing any metal with enhanced naturally occurring radioactivity or cleared from nuclear use, to be informed of this fact and the regulatory framework under which they have been released. Such information should be conveyed with the released materials to the successive suppliers and buyers of the metal scrap - up to and including the melting unit - to allow prior informed approval by the purchaser of the material.

Detection

Material released in accordance with appropriate national regulatory frameworks would have very low levels of radiation that are generally not detectable by commercial equipment used by the scrap recycling and metal producing industries. However, detection capabilities are continuously improving and may alarm with released material: therefore, it is advisable that all incoming scrap material be subjected to the same detection systems.

Reaction

Reactions following detection should be based on the contractual arrangements between seller and buyer.

D.   Controls on the output materials from the metal works – metal, slag and off-gas dust

A final control on the output materials of the metal works, in particular the produced metal, the slag and the off-gas dust should be conducted, thereby providing additional assurance that radioactive materials have not been accidentally introduced into the plant.

It should be recognised that very sensitive detection equipment is needed for the final control of the produced metal because of the dilution that would have occurred when any unnoticed radioactive material was melted with a much larger bulk of clean material. Nevertheless, appropriate measurement of the produced metal will ensure that the final metal product meets the customer's specifications.

CHAPTER I

INTRODUCTION

In order to save as much natural resources as possible, recycling of materials is becoming more and more important: as such, recycling contributes very actively to sustainable development.

This is particularly the case in the metal industry, where recycling has a long history, carried out over a century before sustainable development became an important issue. Compared with other materials, it is a major advantage of metals not just to be recyclable, but to be infinitely recyclable. This advantage has been well recognised since the industrial revolution and presently, about half of the steel production world-wide is made from metallurgical scrap and this percentage is still increasing. This is also the case for most of the other metals.

Although stringent controls are applied to radioactive substances, it can happen that radioactive materials are, under circumstances of loss or lack of regulatory control, introduced accidentally into metallurgical scrap and hence in the metal production process.

Radiation sources, utilising either radioactive materials or radiation generators, are used throughout the world for a wide variety of peaceful purposes, in industry, medicine, research and education, and also in military applications. Many uses involve sealed sources with the radioactive materials firmly contained or bound within a suitable capsule or housing. Some uses also involve radioactive materials in an unsealed form. The risks posed by these sources and materials vary widely, depending on the radionuclides, the forms and activities. Unless breached or leaking, sealed sources present a risk from external radiation exposure only. However, breached or leaking sealed sources, as well as unsealed radioactive materials, may lead to contamination of the environment and the intake of radioactive materials into the human body.

Until the 1950s, only radionuclides of natural origin, particularly radium-226, were generally available. Since then, radionuclides produced artificially in nuclear facilities and accelerators have become widely available, including cobalt-60, strontium-90, caesium-137 and iridium-192. The risks associated with the use of radioactive materials must be minimised and protected against by the application of appropriate radiation safety standards. Regulatory control has, however, sometimes lagged behind developments in the use of radionuclides.

The risks associated with the planned use of radiation sources and radioactive materials are generally well known and the relevant radiation safety requirements generally well established. Nevertheless, accidents can occur during use. In recent years there has been a growing awareness of the potential for such accidents, some accidents having had serious, even fatal, consequences. The attention of the radiation protection community has therefore become focused on the prevention of accidents involving the use of such sources and materials.

More recently still, there has been a growing awareness of the problems associated with radiation sources that for one reason or another are not subject to regulatory control or over which regulatory control has been lost. As radiation sources may be transported across borders, such problems are not necessarily restricted to the State within which the sources were originally used. Such sources are commonly referred to as "orphan sources", a term which is taken here to include:

- ➢ sources that were never subject to regulatory control, but should be under regulatory control as required by the BSS[2];
- ➢ sources that were subject to regulatory control, but have been abandoned, lost or misplaced;
- ➢ sources that were subject to regulatory control, but have been stolen or removed without proper authorisation.

The precise number of such sources in the world is not known, but it is thought to be substantial and possibly is of the order of many thousands. Many sources that are currently regarded as "orphan" were originally used in medicine or industry. Some, however, derive from military activities about which knowledge may not be readily available to the civilian regulatory authorities.

Sealed sources or their containers can have a certain attractiveness because of their appearance or their apparent value as scrap metal. The subsequent handling of such sources and containers by workers and members of the public unaware of the inherent hazards can give rise to external irradiation or, if tampered with, the possibility of internal exposure. This has led to serious injury and in some cases death.

As a consequence, radioactive sources incorporated into metal scrap for subsequent recycling can lead to the contamination of plants and the environment, possibly with serious economic consequences. It may also cause health and safety problems to the personnel during the handling or reprocessing of the scrap, and to the general public during transport and melting of the metal. When inadvertently introduced in the final product, it may additionally have detrimental effects on the marketability of the products and in the worst case on the public health and safety. The economic consequences of the melting of a radioactive source may be extremely high for the concerned economic operator. As a result of international trade in scrap metal, such radioactive material can be transferred from one country to another.

Radioactive material may appear in metallurgical scrap through three different ways:

- ➢ The introduction of discrete radioactive sources into the scrap, due to the fact that such radioactive sources may escape from regulatory control because they are lost, misplaced or stolen. They even may be intentionally abandoned due to the high cost of the management and disposal of such sources.

- ➢ Radioactive contamination of the metal in the process where it has been used, due to contact with either natural radionuclides (for example, in some extraction industry) or man-made radionuclides (for example from the nuclear industry or industrial uses of radionuclides). Such contamination may appear in the scrap stream due to an uncontrolled release of the scrap.

- ➢ Release of material with a very low level of radioactivity, in accordance with a national regulatory framework.

International organisations have issued safety standards and recommendations to keep under strict control radioactive sources or radioactively contaminated materials. Many countries have adopted these safety standards and recommendations in their national regulations. Nevertheless there may be sources over which regulatory control may not exist or the regulatory control has been lost. This is particularly the case in countries where the regulatory and technical infrastructure is not sufficiently developed. Since the metallurgical scrap market is a world-wide market, import of scrap may then cause some additional risk of introduction of radioactive materials into the recycled material.

---

[2] "International Basic Safety Standards for protection against ionising radiation and for the safety of radiation sources", published by the International Atomic Energy Agency under the Safety Series n° 115. This document is jointly sponsored by the Food and Agriculture Organisation of the United Nations, the International Atomic Energy Agency, the International Labour Organisation, the Nuclear Energy Agency of the Organisation for Economic Co-operation and Development, the Pan American Health Organisation and the World Health Organisation.

In addition to regulatory control, the metal recycling and producing industries have organised themselves to reduce the probability of sources that escape the regulatory control being introduced into the recycling loop. Measures are therefore recommended in order to detect such radioactive substances as early as possible. However their detection is not an easy task. Even with the most sensitive and sophisticated equipment, undetected radioactive sources or materials may be introduced into the recycling process. Systematic radioactivity controls have therefore proven necessary on the products leaving the metal works in order to guarantee their quality.

The tolerable radioactivity levels of the recycled products certainly have to satisfy internationally accepted levels established in order to adequately protect the public and consumers. Nevertheless, although these levels are based on stringent public health and safety considerations, for economic reasons resulting from public perception, the metal-works and the down-stream industries do not want to introduce any additional radioactivity into their products in comparison with the typical natural background radioactivity content in the metal. and these industries suggest therefore to use this concept as a voluntary business acceptance limit for metals used in unrestricted products (see § 2.6.2.)

This publication first gives an overview of the present situation regarding the possible introduction of radioactive materials into recyclable metal scrap and their consequences (Chapter 2). Then it describes the prevention measures that should be in place for an adequate control over radioactive materials. The recommendations of relevant international organisations and the actions being undertaken to implement them, are discussed (Chapter 3).

The next section deals with monitoring within the metal recycling and producing industries, in order to detect radioactive sources and material in the metal scrap as early as possible and as much as industrially feasible (Chapter 4). The following section addresses the actions required in case of detection of radioactive materials in the scrap (Chapter 5). These two sections recommend operational practices for the scrap-recycling and metal-producing industry. In case of discovery of orphan sources, arrangements for international cooperation are described (Chapter 6).

Since such monitoring systems may fail to detect the introduction of some radioactive sources or substances, the document further describes a final control on metal works' outputs in order to ensure that the specifications of the products and production residues are effectively met (Chapter 7).

Finally, the publication sets out conclusions and recommendations for the improvement of the management of radiation protection aspects in the recycling of metal scrap (Chapter 8).

CHAPTER II
OVERVIEW OF THE PRESENT SITUATION

2.1. THE METAL RECYCLING LOOP

Metal scrap, as a secondary raw material, has over the last century become as important for the production of metals, as primary ore. As an example, nearly half of all the steel that is produced world-wide is made from scrap and this percentage is increasing. This is a very positive contribution to sustainable development, with particular environmental benefits such as the reduction of emissions during production and the saving of natural resources.

Metal scrap is a sought-after commodity as it provides business opportunities from its collection, through its recovery stage and on to its transport to metal works, rolling mills and foundries world-wide. There is an extended chain of supplier and customer interaction that reaches around the whole recycling loop back to the initial scrap arising. Many economic operators may be involved throughout this loop.

The objective of these metal works, rolling mills and foundries is to deliver products according to the specifications of their customers, including the regulatory specifications. In this respect quality assurance standards clearly state that the producer is responsible for ensuring the compliance with the required specifications of his products. One of these product specifications is the type and activity of radionuclides in the metal.

To ensure the health and safety of the consumers, very stringent quality requirements have to be complied with on the radionuclide content of products.

In practice, consumers of goods do not wish to have any radiation emanating from their purchases. If they were told of a product being radioactively contaminated, they would probably choose alternative products that do not exhibit such characteristics from another supplier. For metal products to be acceptable to the consumers, all producers downstream of the metal works look to the metal works to provide material complying with these wishes. The metal works, in turn, look to their suppliers of secondary raw materials to ensure contractually that the scrap does not contain radioactivity additional to the typical natural background radioactivity content in the metal. However, incidents of companies finding themselves in possession of contaminated scrap continue to be reported. There is a vested interest throughout the scrap cycle not to come into possession of radioactively contaminated metal scrap.

2.2. TYPES OF METAL SCRAP

There are three recognisable types of scrap. These types are the 'home scrap' (synonymously 'own arising' or 'revert scrap'); 'new scrap'; and 'old scrap'.

2.2.1. Home scrap (own arising or revert scrap)

This scrap arises during the production of metals in the metal works itself. The quantity is strongly dependent on the efficiency of the production process and can reach 15% of the total metal production. This scrap does not normally leave the metal works as it is of known composition and can be reused directly for new metal production. It is therefore unlikely that radioactive contamination would occur in this scrap type.

2.2.2. New scrap (traded scrap)

This scrap arises from further processing of semi-finished products e.g., slab, sheet or bar, into the final metal products. Usually it takes no more than one year before this scrap is returned to the primary or secondary metal works. As this scrap arises, it will be of known composition and generally separately retained and segregated for sale into the scrap industry or directly back to the metal works. This scrap would be unlikely to be radioactively contaminated.

2.2.3. Old scrap

After a lifetime which can reach several decades, metal plant, equipment, packaging materials, buildings, engineering construction, cars, ships, trains, aircraft and the like become obsolete. The materials from these goods are sought after for recovery and recycling.

Most of this old scrap metal is not likely to be radioactively contaminated. However, scrap arising from specific activities, such as the nuclear industry or the use of radioactive sources for medical, research or industrial purposes, has the potential to contain lost radioactive sources or to be contaminated in the case of inappropriate regulatory control of these activities. This potential is described in sections 2.4 and 2.5.

Though scrap arisings can be identified from their source as fitting into one of the three categories above, the last two categories may well not remain separated as they pass through the recovery and recycling steps before delivery to the metal works. On the other hand, metal works have over time derived specific scrap quality categories that are suited to their operations dependent on such factors as density, dimensions and composition. It is obvious that the closer the composition of the scrap charged into the furnace is to the final alloyed product specification, the less molten metal refining would be required and the less alloying additions would need to be made.

2.3. BASIC INFORMATION ON RADIOACTIVITY

2.3.1. Radioactivity and radiation

Elements usually exist in one or more isotopic forms. Some isotopes are unstable and undergo spontaneous change, referred to as radioactive disintegration or radioactive decay. Most radioactive isotopes are man-made, but there are a number of naturally occurring radioactive materials (NORM).

Radioactive decay is usually accompanied by the emission of high energy radiation that ionises material in its path. This radiation can consist of alpha or beta particles, gamma rays or neutrons. Each unstable isotope, also called a radionuclide, emits radiation of a characteristic type and energy, which can be used to identify it. Many radionuclides emit more than one type of radiation.

The radioactivity is measured as the number of disintegration's per second. The unit of radioactivity is the "becquerel" (Bq), which represents one disintegration per second. A former unit used was the curie (Ci), which corresponds to $3.7 \cdot 10^{10}$ Bq.

Ionising radiation that reaches the human body may have harmful effects on health, due to the transfer of energy to the cells that the radiation penetrates when going through the tissues. These health effects are measured by the radiation dose received by the body and are expressed in sievert (Sv). One sievert corresponds to a transfer of energy of 1 joule per kg, multiplied by a factor that is related to the biological effects the particular ionising radiation may have on the human cells.

2.3.2. Background radiation

All materials contain traces of radioactive substances, most of which are of natural origin. In addition cosmic radiation reaches the surface of the earth. Together with other sources of natural exposures, the result is that there are always measurable levels of radiation, known as background radiation, in the environment. An order of magnitude of this background radiation is 0.04... 0.1 µSv/h (microsievert/h); however, large variations with place and time occur.

2.3.3. Origin of the radioactivity

There are two main origins of radioactivity. These are:

- Natural radioactivity: this is the decay of radionuclides, which are found naturally in the environment. Such natural radioactivity has been present since the earth's origin.

- Artificial radioactivity: this is the decay of man-made radionuclides. These have been produced since the 1940s.

Both natural radioactivity and artificial radioactivity may be concentrated through human activities to an extent that the risk to human health and to the environment increases.

As with all materials in common use, there are traces of natural radioactivity in metals. The term "additional radioactivity in the metal" used in this document means all the radioactive substances content, which is above the typical natural background radioactive content in the metal.

2.3.4. Discrete sources

For specific purposes, discrete sources of radioactive materials are produced in the form of powder, liquid, pellets or other solid form, which are sealed in a capsule. The capsule or material of a sealed source is strong enough to maintain its containment under the conditions of use and wear for which the source was designed, whilst allowing the emitted radiation to pass through and to be utilised.

Discrete radioactive sources are widely used for industrial, research and medical applications. Their activity could be very high, and therefore the health and safety issues due to the presence of such sources in scrap are of utmost importance.

For storing and transporting radiation sources, and for facilitating their use, a sealed source is usually kept in a container. Source containers have the purpose of shielding the radiation emanating from the source and are therefore heavy and thick walled, made of lead or depleted uranium. Some source containers can be opened on one side to allow radiation to emit as a beam for particular uses. Depleted uranium is often used as shielding material because of its high density. While this uranium is naturally slightly radioactive, the level of this radioactivity is not significant in comparison with the activity of the source inside the shield.

Sealed radioactive sources are normally subject to regulatory control. Those, which escape control, are known as orphan sources (cf. § 3.2.), and these can appear in the scrap stream. It can be very difficult to detect a sealed source, which has been discarded in the scrap metal stream if the shielding remains intact and if the container is surrounded by a large amount of scrap. Despite their strong construction, source containers are vulnerable to rupture by scrap processing equipment whereupon the contents can irradiate and/or contaminate personnel, other scrap and equipment.

2.3.5. Radioactive contamination

Radioactive contamination means the presence of radioactive substances in (called volumetric contamination) or on (called surface contamination) a material or the human body where they are undesirable or could be harmful.

Volumetric contamination

Volumetric contamination of metal appears where the whole volume of the material is radioactive. This may be the result of activation by neutron radiation of the type that occurs in nuclear reactors or accelerators, or may be due to a radioactive source being introduced into molten metal. As a consequence, it is completely mixed throughout the metal and remains so after solidification.

Surface contamination

Surface contaminated metal is likely the most common form of radioactive contamination encountered in the scrap recycling industry. In some cases contamination may be visible as a scale or coating but often nothing can be seen. In practice, this type of contamination can often be removed, for example by chemical or mechanical treatment.

Contamination with naturally occurring radioactive materials comes into this category. This happens when, for example, naturally radioactive materials are deposited as scale in ore processing equipment or in pipes used by the oil and gas industry or by the phosphate industry. Such materials are commonly defined as "NORM" (naturally occurring radioactive materials). When such contaminated equipment becomes obsolete it may be sold into the recycling industry.

Material with surface contamination also appears commonly in the nuclear industry, where contaminated water or air is in contact with pipes and equipment, causing some deposition of radionuclides on the surface of the materials.

There are also components and materials that have been intentionally coated with radioactive materials, such as luminous dials. The use of these radioactive materials may have been exempted or even produced before issuance of regulations. At the end of their lives these components may appear in the scrap stream for metal recovery. Ex-military scrap with such contamination may also appear in the scrap stream.

2.4. THE POTENTIAL INTRODUCTION OF DISCRETE SOURCES IN METAL SCRAP

It is important to know where discrete sources of radioactivity may be introduced in the life cycle of metal products.

2.4.1. Primary and secondary metals production

Metal-industry semi-products, such as slab, ingot or billet, are the result of the mixing, melting, refining and solidifying processes carried out in metal works.

In metal works, a number of material components are mixed and melted down forming a molten bath of metal covered with a layer of slag. The liquid metal is treated, and elements are introduced or extracted, until the melt has the proper specification. The use of metal scrap, which is itself radioactively contaminated, can introduce additional radioactive substances into the molten metal. When the radioactive material is melted in a furnace, the radionuclides are not destroyed and the radioactivity remains. These radionuclides become distributed between the metal, the slag and the off-gas dust, which form in the process. The distribution of the radioactive material between these three phases depends on the chemical and physical properties of the radioisotope involved. Most natural radionuclides pass either to the slag or to the off-gas dust. On the other hand, man-made radionuclides pass to the metal, slag, or off-gas dust, depending on the radioisotope (cf. section 7.1 hereafter).

If the radionuclides were absorbed by the metal, then the semi-products and subsequent products would then also be contaminated. Such instances of contamination have occurred over recent years.

2.4.2. Designers, owners, operators and maintainers of plant and equipment

In designing and constructing plant and equipment, the designer who specifies the need for a radioactive source for measuring, detection or some other radiation purposes, e.g. in hospitals, factories or laboratories, will be aware of the subsequent selection, purchase and installation of the device.

In most countries, intentional uses of radioactive substances are subject to specific regulations and controls by the competent authorities, such as authorisations and inspections, which apply to the whole lifecycle of the source, including its decommissioning and safe disposal.

The last owner or operator of the plant and equipment and those who maintain the plant are in principle also well aware if radioactive sources are being used. At the end of the plant or equipment's life, when the time comes for demolition, the owner has the problem of dealing with any radioactive sources in the facility. The key problem arising at the time the equipment is discarded or the plant demolished is the disposal of discrete sources or radioactively contaminated metallurgical scrap. Such material is expressly not wanted by the recycling industry.

2.4.3. Demolition of plant and discarding of equipment

The demolition sector is largely dependent on the last owner, operator or maintainer of the obsolete plant and equipment for proper information regarding the facility that is to be demolished, removed and transported to the scrap processor.

The proper disposal of radioactive sources, or materials contaminated with radioactive substances (radionuclides of artificial origin or naturally occurring radioactive materials), from obsolete plant or equipment up for sale or removal is naturally not the primary interest of the last owners, especially after many years in service. Although the decommissioning of facilities where radioactive sources were used is subject to regulation and control, it is the loss of knowledge or perhaps the lack of prime interest that makes demolishers vulnerable to handling radioactive sources without their knowledge. A second consideration is the cost of disposal. The final owner and the demolishers both wish to avoid this cost for equipment, which is no longer productive and has value only as scrap. The cost of disposal of the radioactive content might exceed the value of the scrap.

These problems are recognised, and are changing for the better. The following provisions have improved the prevention of the inadvertent intrusion of radioactive sources in scrap.

➢ Controls by regulatory authorities on operators of plants has been strengthened over the years, including decommissioning of sources, thus reducing the likelihood of loss of control over the radioactive sources.
➢ Scrap industry personnel have become more alert to the potential presence of radioactive sources.
➢ Scrap companies are more likely to have detection equipment and will reject the scrap if any discrete source is found.

Despite improvements, the scrap industry sector remains among those most exposed to the risk of receiving radioactive sources from demolition of plants and equipment. A further important consideration is that demolition scrap may be transported in volume directly to metal-works without passing through a scrap-processing site. If detectors are not used at the exit of the demolition site, reliance on detectors at the metal works becomes necessary.

2.4.4. Scrap collectors and processors

The majority of scrap collection companies are very small, with a staff of five or less; however, there are very many companies of this type to be found spread through nearly every town and city. The scrap collectors, who form the largest group by type, in turn supply the smaller number of medium-sized companies. These in turn, supply the even smaller number of large companies. This might be thought of as pyramid-like structure, with a large number of small collectors at the base, who, together with traders, transfer ownership of the scrap up the pyramid until it is finally sold to the last group of large companies that sell direct to the metal works of the world. A chain of contracts exists from the smallest to the largest company. The collected scrap may pass through several stages of sorting and separation before it is gathered together to form the shipment quantities and specific qualities required by the metal works. Most of these scrap companies will certainly have equipment for cutting or shearing metal scrap and many will also have baling presses to compact recovered metals. The larger companies may operate complex integrated plants with many types of scrap-processing equipment. Shredder installations are a good example of the larger specialist plant and can, with their associated media and metal separation plants, be considered a distinct group using specific high technology equipment.

The use of any of these types of equipment could potentially damage sealed and shielded sources, as can the lowest technology use of a hammer and chisel by an unsuspecting workman. However, the risk or probability of such an event happening to an individual company is very small in statistical terms, given the large number of scrap collectors and processors. The pyramidal structure of the scrap industry would suggest that the larger plants would be at more risk of coming across radioactive sources due to the passage of large volumes of scrap through their processes.

When lost or damaged sources appear in the scrap stream, the damage can be severe if irradiation or contamination of personnel occur and if equipment or soil are contaminated. Subsequent clean-up costs can even threaten the viability of the company concerned. Many of the large scrap-processing companies have acquired the means to detect radioactive sources in order to protect their workforce, their customers and their equipment from such risks.

The largest companies which directly supply the world's metal works usually state that their scrap meets internationally recognised specifications or classifications such as the European EFR-EUROFER scrap specifications[3]. This stipulates that "all (scrap) grades shall exclude hazardous radioactive material; material presenting radioactivity in excess of the ambient level of radioactivity and radioactivity in sealed containers even if no significant exterior radioactivity is detectable due to shielding or due to the position of the sealed source in the scrap delivery".

2.5. THE POTENTIAL FOR CONTAMINATION OF SCRAP BY RADIOACTIVELY CONTAMINATED MATERIAL

It is important to know where radioactively contaminated materials may be introduced and produced in the life-cycle of metal products.

2.5.1. Primary and secondary metals production

The use of metal scrap, as opposed to primary ores, in the production process can, if the metallurgical scrap is contaminated, introduce additional radioactivity into the molten metal. If

---

[3] Ref. Eurofer scrap specifications.

this is not detected, the products and by-products can also become contaminated as well as products made with that metal.

2.5.2. Designers, owners, operators and maintainers of plant and equipment

The designer of a nuclear facility will be aware where volumetric or surface contaminated metallurgical scrap is produced during the plant operation. Likewise, in the case of the design and construction of mining, oil and gas facilities, it is also well known where naturally occurring radioactive materials will accumulate and remain in the form of surface contamination.

The owners of mining, oil and gas facilities have to know the level of natural radioactivity appearing during operation and where it is situated in the facility. Likewise, the owners of nuclear facilities such as nuclear power plants, nuclear fuel cycle facilities and other facilities involving the use of radionuclides know the amount of artificial radioactivity used and produced and where it is located in the facility. The key problem is the segregation and appropriate management of the radioactively contaminated metallurgical scrap in accordance with the regulatory requirements.

2.5.3. Demolition of plant and dismantling of equipment

The demolition sector is largely dependent on the last owner, operator or maintainer of the obsolete plant and equipment for proper information on the facility regarding the location and segregation of contaminated metallurgical scrap. It has to be determined beforehand who is responsible for the detection and subsequent proper management of materials contaminated with radioactive substances or naturally occurring radioactive materials.

2.5.4. Scrap collectors and processors

The pyramid-like structure of the recycling industry (described in section 2.4.4.) can be envisaged with the large number of small collectors at the base of the pyramid who together with traders transfer ownership of the scrap up the pyramid until it is finally sold to the last group of large companies that sell directly to the metal works of the world. The pyramidal structure of the scrap industry would suggest that the larger plants would incur greater risk of coming across contaminated materials owing to the passage of large volumes of scrap through their processes.

Contaminated metallurgical scrap may be inadvertently introduced into the scrap recycling processes if the release of these materials is not appropriately controlled. Additionally, materials originating from specific sectors such as the nuclear industry or some extraction industries could introduce low levels of additional radioactivity into the scrap flow. These materials when released in accordance with the national regulatory framework or exempted from any regulatory control would be considered as not harmful, but the perception of the general public could have a negative impact on the marketing of the recycled product.

2.6. PRESENT SITUATION REGARDING LIMITATIONS OF RADIOACTIVE SUBSTANCES IN SCRAP

2.6.1. Regulatory requirements

International organisations such as the IAEA have been working intensively to define radiological standards and good practices with the goal of ensuring the safety and health of the general public, and of the workforce in industries dealing with radioactive substances.

These may be found in IAEA documents such as Basic Safety Standards[4], TECDOC 855[5] and TECDOC 1000[6].

The growing number of obsolete nuclear facilities demands a solution for the relatively large amounts of metal scrap having a very low level of additional radioactivity. Such materials with a very low level of radioactivity may be removed from further regulatory control and released for unrestricted use as normal scrap, if predefined levels of radioactivity are not exceeded. Such levels are defined by the national regulatory authorities to ensure the safety and health of the public. Many countries have clearance levels for mass specific contamination of between 100 Bq/kg and 300 Bq/kg depending on the radionuclides. Some countries have much higher clearance levels.

In addition, it should be noted that surface contaminated objects exceeding 0.4 Bq/cm^2 (0.04 Bq/cm^2 for alpha emitters, other than of low toxicity) fall within the scope of the international regulations for the transport of radioactive material.

2.6.2. Contractual specifications applied in the industry - Business acceptance limits

The criteria set out in 2.6.1. above are essentially based on considerations of the health and safety effects of the residual radiation with respect to the public.

However, in deference to public (consumer) perception, the metal works and down-stream industries only want to use uncontaminated scrap for unrestricted products. (Uncontaminated scrap is scrap without any additional radioactivity beyond the typical natural background radioactivity content found in the metal.) Hence industry would like the setting of a voluntary 'business acceptance limit'.

Such a limit, coupled with practical exchanges of data and a warning system for all partners within the voluntary system, would have the added benefit of preventing the unnecessary rejection of scrap consignments and any unwarranted disposal of materials.

2.7. RESPONSIBILITY IN CASE OF DISCOVERY OF AN ORPHAN SOURCE OR RADIOACTIVELY CONTAMINATED MATERIAL IN METAL SCRAP

The 'Polluter Pays Principle' is used in much environment and waste legislation. It finds a practical application in the 1989 Basel Convention on Transboundary Movements of Hazardous Wastes and their Disposal and in the related Protocol on Liability and Compensation for Damage. The Convention requires the potential polluter to act to prevent pollution and those who cause the pollution to pay for remedying the consequences of that pollution. The "Pollution Pays Principle" is also a component of the European Union policy on the environment, according to Article 174 of the Treaty establishing the European Community. This concept has not yet been used to place the burden of environmentally sound management of the discovered radioactive source/contamination on its generator. In most

---

[4] "International Basic Safety Standards for protection against ionising radiation and for the safety of radiation sources", published by the International Atomic Energy Agency under the Safety Series n° 115. This document is jointly sponsored by the Food and Agriculture Organisation of the United Nations, the International Atomic Energy Agency, the International Labour Organisation, the Nuclear Energy Agency of the Organisation for Economic Co-operation and Development, the Pan American Health Organisation and the World Health Organisation.

[5] IAEA TECDOC 855 "Clearance levels for radionuclides in solid materials" Interim Report for comment – IAEA, Vienna (1996).

[6] IAEA TECDOC 1000 "Clearance of materials resulting from the use of radionuclides in medicine, industry and research" IAEA, Vienna (1998).

countries, the finder of source/contamination instead of the polluter or producer is financially penalised. In the majority of cases the finder is considered liable and responsible for the proper disposal of the radioactive contaminated metallurgical scrap, which was not wanted in the first place.

The regulations for licensing the use of radioactive sources and the occasions where control is lost and sources become orphaned remains of much concern as these 'orphan sources' can do great damage if undetected in the scrap stream.

Voluntary agreements and codes of practice have been drawn up on a national basis, the 'Spanish Protocol' being one example (cf. annex 5). Such codes of practice currently remain limited within the national borders. As further agreements are drawn up and differences between national positions become apparent, the effect could well be to attract contaminated metallurgical scrap to some countries and at the same time to drive contaminated metallurgical scrap out of other countries. This scenario would be unwelcome and can be avoided by harmonising voluntary agreements, codes of practice and legislation.

2.8. PAUCITY OF INFORMATION

The legal burden of correct disposal clearly falls on the owner of the radioactive material and some recent legislation[7] requires that the disposal, recycling or reuse of materials containing radioactive substances be subject to prior authorisation. Despite these requirements, there can sometimes be a lack of information from the last holder, owner, operator or maintainer of plant and equipment that could forewarn recyclers of the potential dangers from radioactively contaminated metals being traded. In this way the burden inappropriately falls on the recycling sector to find the contaminated materials rather than the holder, owner, operator or maintainer to declare such materials. This situation is clearly unsatisfactory and potentially dangerous.

The IAEA maintains a web site of lost and found radioactive sources that are reported to it, but currently there are very few reported. There are undoubtedly many more sources that become lost or stolen and which could potentially enter the recycling chain. Additionally, this web site is presently only accessible to designated national contact points and therefore not accessible to the recycling industry to allow forewarning when radioactive sources are lost.

---

[7] Council Directive of the European Union 96/29/Euratom of 13 May 1996 laying down the basic safety standards for the protection of the health of workers and the general public against the dangers arising from ionising radiation - Official Journal of the European Communities n° L 159, 29.6.1996)

CHAPTER III

PREVENTION OF CONTAMINATION OF
SCRAP THROUGH REGULATORY CONTROL

3.1. GENERAL

Radiation protection aims at ensuring protection against two types of radiation damage:

- Deterministic effects resulting from the killing of cells which, if the dose is large enough, causes sufficient cell loss to impair the function of the tissue. The probability of causing such harm will be zero at small doses, but above some level of dose (the threshold for clinical effect) the probability will increase steeply to unity (100%). Above the threshold, the severity of the harm will increase with dose. Deterministic effects often appear in the case of direct exposure to discrete sources.

- Stochastic effects may result when an irradiated cell is modified rather than killed. Modified somatic cells may subsequently, after a prolonged delay, develop into a cancer. Based on scientific considerations it is assumed that the probability of a cancer resulting from radiation increases with increments of dose with no threshold. However, there is consensus among international organisations that for an individual the probability associated with increments of the order of ten microsievert per year is so low that it does not justify any regulatory effort to further reduce it.

For several decades, international organisations have recommended and industrialised countries have adopted comprehensive legislation on radiation protection and on nuclear safety. This legislation is basically aimed: at the protection of the professionally exposed workers; at the control of radiation sources; at preventing and mitigating the consequences of accidents and on limiting the impact of the operational radioactive material released from installations. The main responsibility for safety and protection is placed on the person using the radioactive substances. Such materials are in general subject to authorisation by the national authorities.

The foundation of the international recommendations as set out by the IAEA for preventing the loss of control over radioactive material, is the regulatory authority, established under national legislation, and empowered:

❏ to issue (or propose to competent bodies) regulations;

❏ to grant authorisations for justified practices such as receipt, possession, import, export, use, transfer and disposal of radioactive materials;

❏ to conduct inspections and enforce regulatory requirements;

❏ to develop the capability to take action leading to recovery or control over radioactive materials in the event of loss, diversion, theft, or unauthorised possession; and,

❏ to implement an enforcement policy to correct non-compliance with regulatory requirements.

In regulations, the main requirement in relation to prevention is that of authorisation for the possession of radioactive materials. Ideally, no transfer, import or export of radioactive materials should take place unless the person making the transfer possesses a copy of the document, issued by the appropriate regulatory authority, authorising the recipient to possess the radioactive materials. However, this procedure is not followed in all countries, especially regarding the export of sources.

Regulations take into account that specialists are needed to ensure radiation protection and whenever there is a risk from ionising radiation, require the involvement of qualified experts whose competence is recognised by the authorities or by appropriate bodies.

Regulations also define the scope to which they apply and to which they do not:

> Radiation exposures, which are not amenable to control, such as exposure from cosmic rays at ground level, or exposure from radionuclides present in the undisturbed earth crust, are excluded from the scope of regulatory control.
> Sources of radiation that could be controlled but for which the effort spent on controls would not be justified by a correspondent reduction of risk which is already considered as trivial may be exempted from regulatory control.

For the latter case, Internationally agreed exemption levels are contained in the International Basic Safety Standards[8]. In a publication of the European Commission[9], the basis for the establishment of such exemption levels is set out. They were derived for moderate scale uses of man-made radionuclides and are a priori not applicable for NORM. Since these levels are expressed both in specific activity and in total activity, they only apply for small amounts of radioactively contaminated materials.

Some examples of such exemption levels for commonly encountered radionuclides are shown in Table 3.1

Table 3.1. Exemption levels of some commonly encountered radionuclides

Radionuclide	Specific activity Bq/g	Total activity kBq
Cobalt 60	10	100
Iridium 192	10	10
Caesium 137	10	10
Radium 226	10	10
Uranium – natural	1	1
Americium 241	1	10

3.2. PREVENTION OF OCCURRENCE OF ORPHAN SOURCES

The risks associated with the planned use of radioactive materials, especially sealed sources, are generally well known. The relevant radiation safety requirements are established on the basis of the general radiation protection legislation. Recommendations are provided in the IAEA Basic Safety Standards[1]. The safe use of radiation sources can be reasonably ensured by the authorities that set requirements on the users of such sources. However, there is a growing awareness of the problem associated with sources that, for one reason or another, are not subject to regulatory control or over

---

[8] "International Basic Safety Standards for protection against ionising radiation and for the safety of radiation sources", published by the International Atomic Energy Agency under the Safety Series n° 115. This document is jointly sponsored by the Food and Agriculture Organisation of the United Nations, the International Atomic Energy Agency, the International Labour Organisation, the Nuclear Energy Agency of the Organisation for Economic Co-operation and Development, the Pan American Health Organisation and the World Health Organisation.

[9] Principles and methods for establishing concentrations and quantities (Exemption values) below which reporting is not required in the European Union - Radiation Protection N° 65, Luxembourg 1993.

which regulatory control has been lost. It is the issue of "orphan sources". An orphan source is defined as follows by the IAEA:

«*A source which poses sufficient radiological hazard to warrant regulatory control, but which is not under regulatory control because it has never been so, or because it has been abandoned, lost, misplaced, stolen or otherwise transferred without proper authorisation*».

The health and economic consequences of possible accidents involving inadequately controlled sources may be particularly severe. A review of such accidents was recently published by the IAEA[10].

Sources, especially sealed sources, are relatively compact and many of them are portable. Therefore problems with such sources are not necessarily restricted to the State within which they were originally used.

In September 2000, the IAEA Member States adopted a Code of conduct on the Safety and Security of Radiation Sources[11] that provides guidances to States for the development and harmonisation of policies, laws and regulations on the safety and security of radiation sources. Full implementation of this Code would significantly help to keep radioactive sources under control. The core of this Code of Conduct is reproduced in annex 6.

3.3. PREVENTION OF RADIOACTIVE CONTAMINATION

The prevention of the contamination of materials that can be reused or recycled in the public domain is based on the establishment of strict regulatory control of the release of materials from regulated activities. This ensures that the management of these materials, including waste, is in compliance with regulatory requirements for the protection of the public and the environment. The basis of these requirements is that the radiological impact of any authorised removal from regulatory control of these materials is sufficiently low as not to warrant any further control. Such removal of materials from regulatory control is called "clearance".

Clearance levels are therefore to be considered as levels below which a particular material may be safely removed from the regulatory system. They are not to be considered as acceptance levels.

The general principles and criteria for clearance have been detailed in the IAEA Basic Safety Standards. They are based on the principles for exemption that were agreed upon in 1988 and on the 1990 Recommendations of the ICRP, which recognises "that the exemption of sources is an important component of the regulatory functions" and iterates the basic criteria, namely that the source gives rise to small individual and collective doses in both normal and accidental conditions and that no reasonable control procedures can achieve significant reductions in individual and collective doses.

The basic radiological criteria for determining which sources and practices within a regulatory control system could be removed from that regulatory control system are as follows:

- the effective dose expected to be incurred by any member of the public due to the practice or use of the source is lower than 10 µSv (microsievert) in a year, and
- either the collective effective dose committed by one year of the practice or use of the source is less than 1 man-sievert (= 1 sievert integrated dose over the whole population subject to the practice and over 1 year), and clearance is the optimum option."

---

[10] IAEA Bulletin, Vol 41 N°3.1999.

[11] "Code of Conduct on the Safety and Security of Radioactive Sources", IAEA/CODEOC/2001, IAEA, Vienna (2001)

The European Commission has recently issued a guidance document[12] on recommended general clearance levels, based on the above criteria. Although these clearance levels are not yet applicable at the international level, some of these are indicated in Table 3.2 as examples of what they could look like in the future.

Table 3.2. Clearance levels of some commonly encountered radionuclides, recommended by a Group of Experts of the European Commission

Radionuclide	Recommended Clearance level Specific activity - Bq/g
Cobalt 60	0.1
Iridium 192	0.1
Caesium 137	1
Radium 226	0.01
Americium 241	0.1

It must be kept in mind that, when the predicted exposure from released materials is not certain to be trivial, appropriate regulatory requirements are maintained, for as long as necessary, to ensure safety.

3.4. MATERIALS WITH NATURALLY OCCURRING RADIOACTIVE MATERIALS

A particular situation may concern naturally occurring radioactive materials (NORM) since some industrial processes, which are usually not under such regulatory control, may concentrate those materials, leading to possible radiological impacts.

Experience with regulation of such natural radiation sources is more limited and the concepts of exemption and clearance relate to human activities in which radioactive materials "are or have been processed in view of their radioactive, fissile or fertile properties"[13]. Exemption and clearance are therefore not directly applicable to such naturally occurring radioactive materials. It is mainly following the recommendations of the International Commission on Radiological Protection (ICRP) issued in 1991 that an effort towards a systematic approach to that problem began to be made.

The European Union has already worked on this particular subject: the approach set out in Directive 96/29/Euratom for such NORM materials consists of three steps:

1) to identify, by means of surveys or by any other appropriate means, the activities which may be of concern;
2) to set up appropriate means for monitoring exposure in the identified activities and, as necessary ;
3) to implement corrective measures (interventions) and apply all or part of the radiation protection system appropriate for activities in which radioactive materials are deliberately used as such.

---

[12] "Practical use of the concepts of clearance and exemption – Part 1 – Guidance on general clearance levels for practices " Radiation Protection n° 122, Luxembourg 2000".

[13] Council Directive of the European Union 96/29/Euratom of 13 May 1996 laying down the basic safety standards for the protection of the health of workers and the general public against the dangers arising from ionising radiation - Official Journal of the European Communities n° L 159, 29.6.1996)

Guidance on the identification of activities in which natural radioactivity may be of concern has been issued by a Group of Experts of the European Commission[14]. Examples of industries for which materials with enhanced concentrations of naturally occurring radionuclides may be of concern are given in table 3.3[15].

As regards exemptions and clearance levels applicable to NORM, guidance is still being developed, taking into account that it would be impractical to define exemption and clearance levels only on the basis of the risk criteria established for man-made radionuclides.

Table 3.3: **Examples of industries for which materials with enhanced concentrations of naturally occurring radionuclides may be of concern**

Industry/product
Phosphate industry (fertiliser production) Phosphoric acid (detergents and food)
Sulphuric acid production
Coal mine de-watering plants
Coal and fly-ash
Metal production : smelters
Magnesium/Thorium alloys
Rare earths : processing of monazite sands, etc.
Foundry sands
Refractors, abrasives and ceramics
Oil/gas industry
TiO2 pigment industry
Thoriated welding rods and gas mantels
Porcelain teeth
Optical industry and glassware

---

[14] "Recommendations for the implementation of Title VII of the European Basic Safety Standards Directive concerning significant increase or exposure due to natural radiation sources" Radiation Protection N° 88, Luxembourg 1997.

[15] "Extract from "Practical use of the concepts of clearance and exemption- Part II, Application of the concepts of exemption and clearance to natural radiation sources" Radiation Protection N° ... (in preparation).

CHAPTER IV

DETECTION AND MONITORING OF RADIOACTIVE MATERIALS IN THE SCRAP WITHIN THE RECYCLING INDUSTRY

4.1. GENERAL APPROACH

The need to monitor scrap for radioactive substances has been recognised since 1983 when radioactive sources were melted at a steel plant in Juarez, Mexico, and at Auburn Steel New York, USA. Detection systems of increasing sophistication have since been developed with the express purpose of monitoring metal scrap for the presence of radioactivity as an additional protection in the event that the regulatory system ensuring control over radioactive materials has failed. Hundreds of millions of tonnes of metal scrap are now monitored annually for radioactivity world-wide. Economic considerations require that these large tonnage's be monitored quickly and in a cost-effective manner. Adequate examination under such conditions is technically challenging, and no existing system can be guaranteed to detect all radioactive sources in scrap. It is also recognised that the fundamental physics of detection makes it unlikely that there ever will be such a system. So, while there is continuing pressure for improvements in the sensitivity of monitoring systems, there is also pressure for multiple monitoring of scrap within the supply chain (see section 4.4) since this can provide a substantial increase in the likelihood of detection.

Most of the technical development for radiation detection in scrap has been undertaken in the steel industry, where the problem has been greatest, and the tonnage of scrap to be examined is greatest. The kind of detection systems used in the steel industry can also be used for other types of scrap. The sensitivity of detection, which can be achieved, is mainly affected by the density of the metal being examined and by the level of the local background radiation. Detection of radioactive materials in aluminium scrap is likely to be easier than in steel scrap, because the density of aluminium is much lower than that of steel (Al 2.7 tonne/m3; Fe 7.8 tonne/m3).

Additionally, visual observation of the scrap during handling may also detect discrete radioactive sources. Indeed, these sources are labelled and marked. Personnel handling the scrap should be educated to recognise such sources. Descriptions of typical sources is included in annex 3.

As such, the probability of contamination of newly produced metal made of recycled scrap could be minimised by an effective defence in-depth approach based on the following hierarchy:

- Prevention of introduction of radioactive sources/contamination in metal scrap by appropriate regulatory requirements and enforcement on users,
- Radiation detection at the entrance/exit of the main scrap yards of the metal recycling industry,
- Visual observation of scrap during successive handling at the different scrap yards during the recycling loop,
- Radiation detection at the entrance of the metal works,
- Radiation detection on the products at the exit of the metal works,
- Radiation detection on the by-products of the fusion (slag, dust).

4.2. FACTORS AFFECTING DETECTION

4.2.1. Radiation penetration

Alpha radiation is not penetrating. It is totally absorbed by a less than one mm of metal or a few mm of air.

Beta radiation is more penetrating. It is, however, absorbed by a few mm of steel.

Gamma radiation is strongly penetrating, and high-energy gamma radiation can pass through tens of millimetres of steel, or hundreds of metres of air. As gamma radiation passes through metal, it is absorbed according to an exponential law, and if there is sufficient thickness of metal, practically no gamma radiation is transmitted. The thickness of a metal needed to completely absorb the radiation depends mainly on the energy of the radiation, and this is a characteristic of the radionuclide which emits it. Thus the gamma radiation emitted from americium 241 has an energy of 60 keV, which is totally absorbed by less than 10 mm of steel, whereas that emitted from caesium 137 has an energy of 662 keV and can penetrate more than 30 mm of steel. An example of the penetration of radiation from caesium 137 through steel is shown in Fig 4.1

Figure 4.1 Absorption of Cs 137 Gamma Radiation by Steel

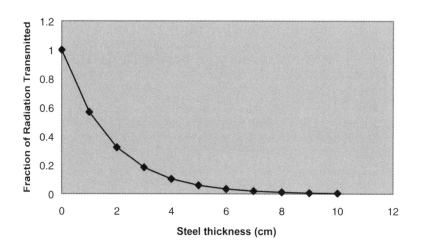

In the practical case of a source in a lorry load of scrap, there is also scattering of the radiation within the scrap. This means that more radiation penetrates compared to the theoretical exponential attenuation example shown in Fig. 4.1 making the detection of a source a little easier.

Neutron radiation is very strongly penetrating. It can pass through hundreds of millimetres of metal or hundreds of metres of air. It usually has gamma radiation associated with it. Additionally, neutron radiation can activate certain materials.

4.2.2. Background radiation

Levels of background radiation are typically in the region of 0.1 to 0.2 microsievert per hour (μSv/h) but can be much higher at some sites (for example, in the neighbourhood of granite geologic formations). Detection of radiation emitted from metal scrap has to be distinguished from this varying background, which can be considered as noise in the detector system.

The practical consequence is that there is a limit on the sensitivity of detection that can be achieved when monitoring for radioactivity in metal scrap.

4.2.3. Methods of radiation detection

Geiger counters.

A Geiger counter has a gas-filled detector. The passage of radiation through the detector causes ionisation of the gas, and allows a pulse of electrical current to pass through it. Electrical circuitry registers the pulses and this allows the level of radiation to be determined by the instrument. Geiger counters are satisfactory for use in radiation safety work, but they have a low efficiency of detection, and are not the method of choice for the monitoring of metal scrap for radioactivity.

Scintillation counters.

Scintillation counters have a detector containing a material, which emits light when struck by ionising radiation. The scintillant material is often either a special plastic material, or an inorganic crystal such as sodium iodide. The light is detected and the signal amplified by a photomultiplier which emits an electrical pulse. Electrical circuitry registers the pulses and this allows the level of radiation to be determined by the instrument. Scintillation counters have a high efficiency of detecting radiation and are the method of choice for monitoring metal scrap for radioactivity.

Additional electronic equipment on scintillation counters may discriminate the amplitude of the impulses, so as to be able to determine not only the overall activity, but also individual radionuclides and their activity separately (such equipment is commonly called a spectrometer).

4.3. GENERAL CONSIDERATIONS WHEN MONITORING METAL SCRAP FOR RADIOACTIVITY

All systems used for the monitoring of tonnage amounts of metallurgical scrap for radioactivity rely on the detection of the gamma radiation that is emitted from the scrap. Alpha radiation and beta radiation are too readily absorbed by metal scrap, and cannot be detected. It is often the case, however, that a radioactive material emitting alpha or beta radiation also emits detectable amounts of gamma radiation. An example of this is naturally occurring radioactive material (NORM), which generally emits alpha, beta and gamma radiation.

Although gamma radiation can penetrate metal, a proportion of the radiation is absorbed by the metal. The amount of absorption increases exponentially with the thickness of metal, so that for large thicknesses of metal negligible amounts of radiation escape (see Fig. 4.1). When radioactive sources are in use they are often kept in heavy metal housings, which absorb much of the radiation and so make handling of the source safer. Because nearly all of the radiation is absorbed by the housing, and little radiation is emitted, detection of a source in a housing in the scrap is difficult.

Almost total absorption of the radiation can also occur in metallurgical scrap, but because a scrap load is partly metal and partly air, the distance the radiation can penetrate through scrap is several times greater than the distance it travels for solid metal. Trials of detector systems in the United States of America examined the practical limit to the depth of scrap, which can be monitored effectively for radioactivity. For the trials a shielded source of moderate size was buried in steel scrap. The results showed that the probability of detection of the source decreased greatly as the thickness of scrap between the source of radiation and the detector increased. For a thickness of scrap of 0.5 m, the probability of detection of the used source was of the order of 10%[16]. To maximise the probability of detection it is clearly preferable therefore, to perform the monitoring on small amounts of scrap. It is also clear that direct monitoring of very large tonnages of scrap, such as those contained in a ship, will not be effective, and the monitoring would have to be performed as the ship is being unloaded.

Once a discrete radioactive source has been found in scrap and removed, the remainder of the scrap having been monitored is usually free for normal use. Special care is needed, however, if the scrap is surface contaminated or contains radioactive substances in a dusty form which can be scattered around, and can be inhaled or ingested.

4.4. MONITORING POINTS

Scrap is usually handled and transported several times before it is used, and so there are a number of opportunities to monitor it for radioactivity. The usual monitoring points for road and rail traffic are at the entrance to the scrap yard or to the melting plant. These points are convenient, since all the scrap entering the site is examined, and the origin of the material is known.

There is also some monitoring of traffic at national borders. The effectiveness of this depends on whether there is specific monitoring of metal scrap, or whether the monitoring is of all traffic. The detection systems used for the monitoring of all traffic are less sensitive than those used for the monitoring of metal scrap, and are not an adequate substitute for the specific monitoring of scrap.

The disadvantage of monitoring scrap loads in vehicles is that the scrap is in large amounts (20 tonnes or more) and the time available for monitoring is usually short. The result is that the sensitivity of detection is limited, for the reasons explained above. This limitation can be overcome to some extent by monitoring scrap loads two or more times within the transport chain, with mixing of the scrap in between the monitoring events. For example, if the scrap is first monitored at the entrance of a scrap yard, where it is likely to be unloaded, sorted and re-loaded, and then is monitored on exit, the overall possibility of detecting radioactivity is substantially increased. It follows that it is in the interest of the final user of the scrap to monitor scrap as it arrives at the site, and where possible to purchase scrap from suppliers who perform monitoring themselves.

Within scrap yards and scrap melting plants, the scrap is often handled by grab in small amounts that offer the possibility of greater sensitivity of detection. The disadvantages of grab detection systems are that once the scrap is unloaded from the truck or the wagon, the transfer of ownership has taken place, and, because of the working environment, the detector is vulnerable to damage; in addition a single site might have a number of grabs, and to monitor all the scrap, a detector has to be fitted to each of the grabs. There are currently monitoring systems commercially available which are designed to be attached to grabs, and these are becoming more widely used.

Attaching detectors to magnet cranes is more difficult than for grabs. Both very strong magnetic fields and the elevated operating temperature of the magnet can compromise the reliability of any detector system attached directly to the magnet.

4.5. FIXED MONITORING SYSTEMS

4.5.1. General features

Scrap is routinely monitored using large fixed systems that are sometimes known as portal monitors. These systems check incoming vehicle loads of scrap in less than 30 seconds, without the need for an operator. There are a number of manufacturers of these systems, but nearly all systems share some general features.

---

[16] LaMastra A., "Advances in monitoring scrap steel for radioactivity", Iron and Steel Engineer, May 1999, 48-50.

4.5.2. Detection of the presence of a vehicle and speed measurement

The system detects the presence of vehicles so that the signal from background radiation can be distinguished from that measured when a vehicle is present. Detection of the vehicle is usually done by photocells, which monitor it while it is passing through the detector system. It is important that the vehicle moves slowly through the system, so that maximum sensitivity of the radiation detector can be achieved. The systems therefore often incorporate a measurement of the speed of the vehicle, and some send an alarm if the speed is too high. The manufacturer of the detection system will give advice on the maximum speed for the vehicle but a typical maximum speed is 5 km/h. It is suggested that the manufacturer's recommendations are followed since the maximum speed is an important factor in ensuring that adequate monitoring is achieved.

4.5.3. Detector heads

The radiation from the scrap is detected using well–established scintillation devices (see § 4.2.3. above). In a typical arrangement there is one set of detectors on each side of the roadway, and the scrap load passes between them. It is clear from the study mentioned[17] that such a system will have limitations with respect to the depth of scrap in which radioactive sources can be detected. Having additional detectors above and below the scrap load can enhance the sensitivity of detection.

The detector array is adjusted to respond to the range of gamma ray energies of interest. It is recommended that the lowest energy which can be detected should be not higher than 50 keV. This ensures that it is possible to detect Americium 241 (gamma emission 60 keV), which is a radionuclide in common use. There is little benefit in endeavouring to detect lower energies because these are almost wholly absorbed within the scrap load, and in addition the radiation background at these lower energies is very high, making sensitivity of detection very poor.

It is recommended that the highest energy which can be detected should be not less than 1400 keV, to ensure that it is possible to detect cobalt 60 (gamma emissions 1173 and 1332 keV) which is the highest energy radionuclide in common use.

The detector systems are usually in permanent operation, either gathering data on the background radiation or on the scrap loads as they pass through the system. The radiation profiles obtained from a detector system during typical situations are shown in Figures 4.2, 4.3 and 4.4.

Figure 4.2 shows the typical change in signal as a vehicle carrying the scrap load passes without any abnormal radioactivity. At the approach of the vehicle the level of background radiation declines because the scrap load shields the detector system from the background radiation in the area. As the scrap load exits from the system the level of radiation increases back to the higher level. The decrease caused by the scrap load is larger for a large tonnage of scrap, than for a small tonnage of scrap (cf. fig. 4.2).

---

[17] LaMastra A., "Advances in monitoring scrap steel for radioactivity", Iron and Steel Engineer, May 1999, 48-50.

Figure 4.2 Radiation profile of large and small loads

Radiation levels observed when monitoring scrap for radioactivity: no abnormal radiation present

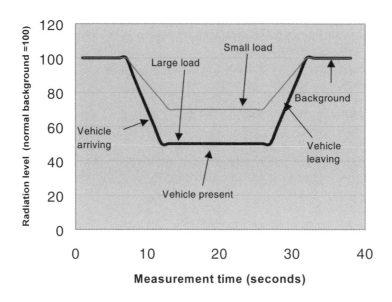

Figures 4.3 and 4.4 show the typical changes in signal when the scrap contains a region with a level of radiation above background (fig. 4.3) and with a relatively low level of radiation (fig. 4.4).

Figure 4.3 Scrap containing a region with a level of radiation above background

Radiation levels observed when monitoring scrap for radioactivity: level of radiation above background detected in scrap

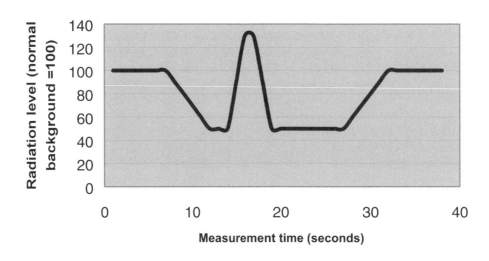

Figure 4.4 Scrap containing region with a low level of radiation

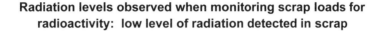

Radiation levels observed when monitoring scrap loads for radioactivity: low level of radiation detected in scrap

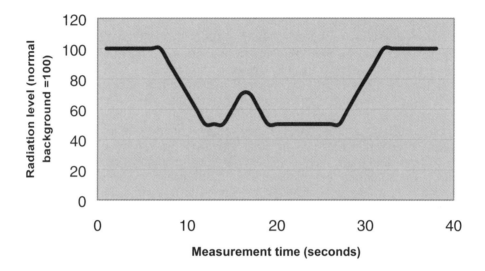

In the example shown in fig 4.3, the level of radiation exceeds the background when there is no vehicle present by 30%. It is nonetheless a very low level of radiation, and sensitive detection equipment is needed to identify the presence of this abnormal level of radiation in the short time available.

4.5.4. Computing hardware/software

The signals from the detectors pass to a computer where the data are analysed. The reference value for the local background radiation is constantly updated with readings received when no vehicle is present. In this way the sensitivity of the system is maximised with respect to the level of background radiation and the number of false alarms is minimised. The signal obtained when a vehicle is present is compared with the current value for the background, and a decision is made on whether radioactivity is present in the load. The setting of the alarm threshold is crucial to the sensitivity of the system. The highest possible sensitivity of detectors (around 5 nSv/h) is much higher than the background radiation level in the environment and even much higher than the background radiation level with a truck passing in front of the detector; therefore, the alarm threshold has to be set so as to ensure a very low rate of false alarms (< 0,1%). These needs have led manufacturers to develop specialised methods of analysing data and identifying the presence of abnormal levels of radiation.

4.5.5. Alarm criteria and false alarms

Figures 4.2, 4.3 and 4.4 show the main situations experienced by a detector system. Figure 4.2 shows the data from a large scrap load containing no abnormal radioactivity and from a small scrap load containing no abnormal radioactivity. Neither of these loads should cause the system to alarm. In practice a sensitive alarm system will sometimes give false alarms, but these should be at a very low rate. A practical false alarm rate would not exceed 0.1% (i.e. one load in a thousand) when clean metallic scrap is being monitored. Asking the truck to repeat the check in case of an alarm, to confirm this measurement can identify false alarms.

Fig. 4.3 shows a situation where the detector system has found a level of radiation from one part of the scrap load that exceeds the level of background radiation by 30%. It is reasonable to be expected that the detector system should be able to trigger an alarm reliably in this situation.

Fig. 4.4 shows a situation where the detector system has responded to a level of radiation from one part of the scrap load that exceeds the level of background radiation with the scrap load present by 30%. A sophisticated detector system is able to distinguish the abnormal level of radiation and trigger an alarm.

Detector systems are designed primarily to monitor scrap for radioactivity, but because the monitoring systems are most usefully placed at the entrance to a site, other materials may also pass through them. A sensitive detector system that is designed to monitor metal scrap is likely to register an alarm on some of these other materials. These alarms occur because the alarm threshold of the system is set for metal scrap, which has a very low typical natural background radioactivity, whereas many other common materials are more naturally radioactive. A common example at steel plants is the refractory bricks used to line the furnaces. These materials have sufficient naturally radioactive content to cause the system to alarm. Such alarms are inevitable on a sensitive monitoring system and can be disregarded if there is confidence that the cause is known. Different alarm settings could help to solve this problem. These alarms are not regarded as false alarms since they are correct detection of the slightly higher level of radiation in the material. They are sometimes called "innocent alarms". Some authors nonetheless report them as false alarms since they are not detection of radioactive substances in scrap. It is recommended that the term 'false alarm' be reserved for events in which an alarm has been registered, on metal scrap, or other material, but for which no corresponding source of radiation can be found by searching of the load.

The technology for detection of radioactive substances in metal scrap, using large fixed monitoring systems, is not able to detect all types of radioactive sources under all practical circumstances. There is therefore pressure to improve the systems, and to achieve increased sensitivity of detection. Manufacturers have different methods of achieving high sensitivity, and not all have the same level of sensitivity. Some systems have sophisticated software that assists in processing the signals and deciding whether radioactive material is present. It is therefore inappropriate to specify all the details of the detection equipment used. It is preferable to specify the sensitivity of detection, which can be achieved, although it is also very difficult to specify a reproducible test for sensitivity (see § 4.5.7. below).

4.5.6. Peripheral equipment

A complete monitoring system can have equipment associated with it such as methods for recording data, printers/visual displays, audible alarms to show whether radioactivity is present and whether a vehicle has exceeded the allowed maximum speed for the monitoring system, traffic lights to control the entry and exit of vehicles, and barriers to prevent vehicles leaving before measurement is complete.

4.5.7. Practical testing of system performance

4.5.7.1. Type testing

The ultimate test of the performance of a system is its ability to detect abnormal levels of radiation in scrap loads. It is usual therefore to test a system by a trial in which small radioactive sources are placed in typical scrap loads. The tests can include the ability of the system to detect sources contained in safety shielding of the kind typically used to hold radioactive sources. Such tests provide useful information, but are time-consuming and costly. In addition they are difficult to repeat in a reproducible manner because each load of scrap is different, there is a wide range of vehicle dimensions, and factors, such as the background radiation, which vary according to time and place. Hence it is difficult to compare the performance of different detector systems unless they are tested under the same conditions with the same radioactive sources and the same load of scrap. The Steel Manufacturers Association of the USA has arranged such tests, and a number of manufacturers of detector systems have participated[18].

The difficulty of specifying a comprehensive, realistic and reproducible test means that there is no established standard for testing the performance of detector systems of this kind.

In general, manufacturers do supply some indication of the sensitivity of detection, which can be expected under specified test conditions.

Simple tests have been suggested which are largely independent of the scrap quality and are therefore reproducible from site to site. The tests involve the detection of radioactive sources having an activity of 3.7 megabecquerel (MBq) (100 microcurie) or less. Use of such sources allows a practical test to be performed without significant radiation hazard to those involved. In these tests, the effect of attenuation by scrap is mimicked by placing a small radioactive source behind steel of a specified thickness.

    i)   A test source of 3.7 MBq of Caesium 137 is placed in a steel container of 110 mm diameter and with 50 mm minimum wall thickness. This test source is placed inside the wall of the vehicle carrying a normal scrap load. This arrangement is driven through the detector system to check its performance. Such a test resembles the real situation.

    ii)   A test source of 0.37 MBq of Caesium 137 is mounted behind a steel plate of 5 mm thickness on the inside wall of a vehicle carrying a normal scrap load. This arrangement is driven through the detector system to check its performance.

Test (i) is more realistic than test (ii) but is more difficult to arrange. Tests of this kind do have some limitations compared with actual tests of radioactive sources buried in scrap but they also have great potential benefits as the basis of a test, which can be reproduced reliably.

The tests in which the source is placed inside the body will not show identical results even for the same detection systems tested at different times and different places. Moreover, using sources of the same dose rate gives no way of obtaining the quantitative characteristic of the system allowing its sensitivity to be detected. This prevents comparing the characteristics listed in the manufacturer's documentation with values of the system's sensitivity in actual practice and also prevents comparison with the control results. Therefore there is a good reason to add test (iii) which makes it possible to evaluate the sensitivity of a system in quantitative values by determining the minimum increment of the radiation dose rate on the surface of a detector which can be detected by the system with a minimum probability of 0.95. This also provides the possibility of comparing the system with the control results.

    (iii)   A source of 0.37 MBq is mounted on the outside of a vehicle. The vehicle with this source passes the detection system 8-10 times: this allows the determination of the detection probability of this source, but it will not introduce any loss of radiation energy and it will allow primary photons to strike the detector directly.

        The above procedure is repeated by mounting sources of lower activities (lower dose rates) on the outside of the vehicle, whereupon the detection probability curve versus the activity of the source is constructed. The

---

[18] LaMastra, "Advances in monitoring scrap steel for radioactivity", Iron and Steel Engineer, May 1999, 48-50.

sensitivity of the control system is determined as the minimum increment of the radiation dose rate on the surface of the detector, as detected by the system with a probability of >0.95.

On trials of detection systems, it is necessary to also determine the possibility of detection of a source of 0.37 MBq ^{241}Am at the distance of 0.5 m from the detector. This allows the lower limit of the energy range of detectors to be determined

4.5.7.2. In-service testing

Once a detector system has been installed it is necessary to check that its performance has not changed. Simple testing methods can be used, and the most usual procedure is to measure the response of the system when a very small radioactive source is placed at measured distances from the detectors. The tests typically involve the detection of radioactive sources having a strength of less than 3.7 MBq (100 microcuries). Use of such sources allows a practical test to be performed without significant radiation hazard to those involved. Testing intervals are usually advised by the supplier of the system, but are typically three months. Full checking of a system will include an examination of the records of past measurements and alarms.

Customers who purchase scrap from the site might wish to audit the performance of the detector system. They will require proof that the system has been in constant operation, and that all scrap loads have been checked. Adequate record keeping is therefore essential.

4.5.8. Reliability

It is desirable that systems have a high level of reliability since all scrap must be checked. It is advisable to have systems which are self-checking and give an alarm if there is any failure of the system. On some sites duplicate monitoring systems are installed so that a failure of one system does not stop the flow of scrap. The system needs to be robust with respect to disturbances of the electrical supply, and to electromagnetic interference, both of which are commonly experienced in metal-producing plants. It is helpful if detectors are able to re-start automatically after interruption of the power supply. Since they are typically in an outdoor environment, it makes sense that the detectors remain functional over the temperature range likely to be experienced at the installed location.

4.6. PORTABLE DETECTOR SYSTEMS

There are a number of manufacturers of hand-held radiation detectors, and a wide range of instruments are available. These instruments are often used in radiation safety work, and are usually calibrated to recognised standards. Alpha, beta, gamma, and neutron radiation can be monitored by use of the appropriate instrument. Such instruments have several roles in checking scrap for radioactivity:

-       At the point of origin of scrap, hand-held monitors can be used to check individual items, which are suspected of containing radioactive substances. Radioactive items can then be separated from the scrap stream at source.

-       At scrap handling sites or melting plants which use only a small quantity of scrap, and where the installation of a large fixed system is not justified, it is possible to use a hand-held instrument to check incoming scrap for radioactivity.

When alarms are given by installed monitoring systems, a hand-held unit is used:

1.   to check whether or not the levels of radiation in the area are safe. Installed monitoring systems do not give sufficient information for this purpose,

2. to identify the area of a load which might have been the cause of the alarm,
3. to search the scrap from a load and identify the source of the radiation,
4. to examine any radioactive items that may be isolated from the load and determine the degree of hazard they present e.g. levels of gamma radiation, and the presence of alpha radiation.

When a scrap load has caused an alarm on a fixed detector (which is not a false alarm), it is because of the level of gamma radiation in the load. The initial stage in searching for the cause of the alarm is therefore the use of a hand held gamma radiation detector. Instruments used to search scrap for radioactivity need to have a high sensitivity of detection, so a unit using a scintillation detector is more suitable than one using a Geiger counter. The instrument selected needs to be capable of identifying levels of radiation which exceed the level of background by < 30% in a practical situation when used by a trained operator.

The first stage of the inspection should always be to ensure that the level of radioactivity is not dangerously high. If the levels are high, specialist assistance might be needed. If the levels are a negligible safety hazard, then a more detailed search of the scrap for the source of the radiation can be made. Once the source of the radiation has been found, the nature of the radioactivity can be defined in more detail using monitors for alpha, beta, and neutron radiation. Specialist assistance will be needed to ensure the safe handling and disposal of the material which has been found. In most cases once the source of radiation has been found and removed, the remainder of the scrap can be safely used in the normal way.

4.7. VISUAL OBSERVATION OF THE SCRAP

Radiation cannot be detected by visual observation, but visual observation of the scrap may indicate the presence of radioactive materials in it.

This is particularly the case for discrete sources, which are contained in thick walled shielding, and, when in use, the containers of these sources have clear indications of their content (among others, the radiation symbol). Training of the workers handling metal scrap in the recycling industry, in recognising such indications, and in recognising the different types of container used for storage and transportation of radioactive discrete sources may help to detect radioactive sources. Annex 3 gives a description of several types of discrete sources commonly used in industry and in medical applications. However, one has to be aware that the indications on the containers may have disappeared during handling of the scrap and that the external appearance of the containers themselves may have changed as a result of the processing of the scrap at the scrap yards.

Also, the origin of the scrap may help to detect possible radioactivity content. If the scrap is originating from industries where naturally occurring radioactive materials can concentrate, such as in the extraction industry or the phosphate industry, the probability of the presence of radioactivity is enhanced, and more careful control is desirable. This is also the case for scrap originating from specific activities such as the nuclear industry or from places that use radioactive sources for medical or research purposes.

Such observations have to be made as early as possible in the supply chain of the scrap. Their effectiveness is much increased when they carried out as close to the origin of the scrap as possible.

Adequate information and education of the personnel at the scrap yards is therefore recommended.

CHAPTER V

ACTIONS TO RESPOND TO A DETECTION OF RADIOACTIVE MATERIALS IN THE SCRAP WITHIN AND OUTSIDE THE RECYCLING INDUSTRY

5.1. RESPONSIBILITIES OF THE PARTIES

5.1.1. General description of scrap flow and economic operators in scrap trading

The metals recycling industry has been fully described in Chapter 2. It can be envisaged as a flow process. The points of origin of the flow are very numerous and may be very low-volume. They include both large and small-scale operations, from the dismantling of industrial structures to the collection of obsolete household products. The main sources of materials are from industry, transport, general engineering, electrical engineering, building and construction fields as well as arisings from commercial premises and households. The important initial processes carried out in the recycling chain are that of collection (concentration of materials) and of sorting (or mechanical processing if necessary) followed by the transport of the materials to fewer, larger merchants handling greater volumes of materials (who may also further sort and process materials). Part of the material is traded internationally. In all cases remelting operations are the ultimate destination of the recovered materials. At various stages throughout this flow, ownership of the metals might change, and at some stages, the flow might be checked for radioactivity.

5.1.2. Determination of the ownership of scrap during trading and transport

Contractually, at all stages of the scrap flow, ownership of the metals is usually clear. There is a seller and a buyer, with a contract between them. There may be one or more intermediary carriers to undertake the necessary transport, but ownership will simply pass from seller to buyer at a contractually defined point.

In general trade specifications and classifications, there are specific clauses that state that the scrap must be free of radioactive contamination. There is no intention by the scrap trade to purchase radioactively contaminated metal. In general, contracts will also specifically exclude it, separately to any reference to the trading specifications and classifications, and radioactively contaminated material when detected will remain the seller's property.

Once radioactivity has been detected, tracing the ownership of this radioactive material therefore requires an inquiry against the flow of the material from the buyer of the scrap to the previous seller, and so on through the 'chain of contracts' until it reaches the originator.

In the determination of the ownership of radioactively contaminated material outside of the contract trail, much depends on the nature of the contaminated metal. If it is undifferentiated and mixed due to earlier sorting processes, the route may not in practice be traceable. Equally, the contaminated metal may be identifiable pipework with naturally occurring radioactive materials (NORM), or a gauge, meter or sealed source with identification or certification that makes it traceable. Even though no contractual link may exist between a sealed sources last owner and the contractor who demolished that company's plant, identification may still be possible. However the trail may lead back to a defunct company.

Problems may also arise owing to the point of transfer of ownership being different to the point of detection of the radioactive materials. The point of transfer will generally be on delivery by the seller, as defined in a more or less detailed contract. A simple, short contract will suffice for a truckload at a local level, but for larger volumes a comprehensive contract will detail the seller's delivery obligations, and include the buyer's obligations to facilitate that delivery. For international trade, the contract will generally be based on Incoterms 2000[19]. Where the two parties have a continuing commercial relationship, they may rely on established practice to guide them.

Owing to the relatively high risk of an accident during loading and unloading, any well-drafted contract is explicit as to precisely where risk transfers from seller to buyer. For example Incoterms FCA (Free carrier)[13] covers delivery by seller to buyer's nominated carrier at a named place. Risk in loading then lies with the seller, but not risk in unloading.

When considering the possibility of radioactive metals being present, there may be conflict between delivery as defined by contract, and the point of detection. If the buyer's carrier uses a truck, loaded by the seller on the seller's premises, and the truck delivers to the buyer's premises FCA, it is quite clear that risk is transferred to the buyer once the truck is loaded. However, as suggested in 5.1.1, many transfers of recycled metal will involve a smaller-volume seller and a larger-volume buyer. A small volume seller may not check the metals for radioactivity at all because it is beyond the seller's capacity, or is economically impracticable to do so. At best, small and medium-sized enterprises (SMEs) may use less sensitive equipment. The contamination may then first be detected while passing through the large volume buyer's radiation detectors on the buyer's premises, by which time risk has passed to the buyer. Insofar as a risk to seller may be revealed by a detection system operated by the buyer, there is a duty for the buyer to explain how and why the load was rejected.

The foregoing discussion also assumes that there is a universally accepted definition of what is acceptable as far as radioactive content is concerned.

➢ Metals contain some typical natural background radioactivity, as do all other associated materials.

➢ The regulations of individual countries related to the radioactive content of commoditiies may vary widely, and material deemed to be at a level below regulatory concern in one country might set off alarms in another. As described in Chapter 2, internationally accepted clearance levels are developed by the IAEA in order to avoid such differences in regulations between countries.

➢ Within the industry, acceptance criteria for radioactivity content may also differ in the area between 'below regulatory concern' and 'typical natural background radiation level'.

Another difficulty is the fact that acceptance criteria can be expressed in terms of dose rate or radioactivity concentrations, which differ from measurements made in practice. Correlation between a radioactivity concentration and a measurement made on a scrap detector is not simple and varies according to many factors. Measurements are also made more difficult by the presence of natural background radiation. It may often be possible to encounter background radiation levels at the site far greater than a country's regulatory limit.

5.1.3. Responsibility of operators of detection equipment

Recycling companies that have invested in detection equipment will have trained personnel to operate the equipment. These and other companies may be able to take advantage of radiation safety awareness courses for their managers and supervisors, and operators, provided for by their national associations. International associations, such as the Bureau of International Recycling (BIR) also provide educational publications to member companies, as do many national associations.

---

[19] INCOTERMS 2000.

The most limited view of this responsibility would be that of safeguarding the operator's staff and premises, so that any suspect shipment is rejected immediately on detection, and the carrier instructed to return the shipment to the originator without further investigation. This neglects the wider public duty of the operator. Best practice will involve investigating every alarm, so as to determine its cause and isolate any radioactive material discovered in a safe manner. When the seller has no detection equipment, or less sophisticated equipment, the operator could be seen to have a responsibility of explaining to the seller why the scrap was rejected (i.e. the detection equipment and methods used, and results). The large volume buyer incurs substantial costs, not only in procurement and maintenance of the detection equipment, administration costs and in training of staff to respond to alarms, but also in storage and disposal of such radioactive metals as may be found. Ideally, the concept of the "Polluter Pays" principle should be used for the management of the discovered radioactive sources and radioactive contamination.

The operator of detection equipment is also unwittingly exposed to increased costs. This is particularly true if somewhere in the flow process there has been wilful concealment of known radioactively contaminated material. The operator of detection equipment can never exclude this possibility, remote though it may be.

It may fairly be argued that Governments should recognise this contribution to environmental protection, but while that case is being pursued, the responsibility of the operator must be fully discharged, as described in more detail below.

5.2. ACTIONS TO BE TAKEN WHEN AN ALARM LEVEL HAS BEEN EXCEEDED

5.2.1. Immediate actions to be taken when an alarm level has been exceeded

These may be summarised as follows:
- Repeat measure to check alarm validity ;
- Isolate the vehicle or container ;
- Seek expert assistance ;
- Carry out actions as required by national laws ;
- Use handheld detectors to locate and characterise the activity, under the responsibility of the seller of the scrap ;
- Contain and possibly shield the radioactive material.

All the foregoing actions should be well understood by the operator's personnel, and be recorded as written management instructions for their reference.

Typical portal detection systems operate by detecting variations above the natural background level rather than measuring the amount of activity in absolute terms. Since the background fluctuates, both in time and between different geographical areas, alarms as a result of such fluctuations are possible. The first step is to repeat the movement of the container through the detection system several times, and confirm that the alarm is genuine. Although unable to distinguish between a small, unshielded source, and a large, well-shielded one, the detection system will also give a useful indication as to the scale of the hazard, particularly whether it is localised or distributed, and how closely it is safe to approach.

After a confirmed alarm, the container or vehicle should be isolated as expeditiously as possible within a quarantine area with limited access (inside or outside the site of the factory), to be accessed by suitably trained and equipped personnel only.

At this point, the skills of a Radiation Protection Advisor are required. Smaller operations may need to contact an external source for the necessary expertise, while a larger operation may have the skill in-house. It is dangerous to approach a radioactive source of unknown strength without the proper monitoring equipment in the hands of someone trained to interpret the readings. The following suggested actions assume low levels of radiation, and in rare cases more radical measures might be appropriate, not excluding evacuation of a localised area surrounding the container.

All work carried out local to the activity from the point of alarm onwards entails potentially incurring a dose by the worker, and that this must be justifiable so as to keep any dose as low as reasonably achievable. The health risks are discussed further in 5.2.2. below.

In most cases, the external levels of radiation will not prevent use of handheld monitors to pin down more precisely the position of the radioactive materials within the load and to characterise it by type and quantity. It may be possible to start unloading, and locate the radioactive materials visually by reference to radioactivity warning signs, recognisable gauges, meters, or other particular components. It is not unknown for the container/vehicle itself to be the problem. Immediate containment of the radioactive materials is important. With the radioactive substances contained and local radiation levels checked safe, the situation is under control. Containment methods are discussed in 5.3. below.

5.2.2. Determination of risk to human health or the environment

Radioactive material could pose both an internal and external health risk. The internal risk arises whenever there is loose radioactive material contamination in that it is possible to inhale or ingest the material itself. If the radioactive material is in a sealed source, then there is only an external hazard from direct exposure to radiation emitted from the source. Sealed sources have been the cause of some serious injuries and even death to persons attracted to these shiny metallic objects. At the levels of surface or volume contamination likely to be associated with recycled metals, the health effects are longer term (stochastic effects – cf. § 3.1.).

Risks from direct radiation exposure can be limited by one or more of three actions. These are: a) reducing exposure time; b) moving away from the source, and c) placing shielding material around the source. To take these actions requires some measuring equipment because the human body cannot sense ionising radiation. We do not sense the background radiation to which we are continuously exposed. Exposure to say one hundred times background if only occurring once, for one hour, in a year, does not increase the annual dose significantly. Since the normal background is in the order of 0.1 to 0.2 µSv/h (microsievert per hour), and since the limit for any individual of the public is set at 1 mSv/y (millisievert per year), levels of tens of microsievert per hour are to be treated seriously.

In the case of surface contamination, inhalation and ingestion are readily prevented by simple precautions, appropriate education and the normal personal protective equipment. It is particularly important to wear gloves and face masks when necessary, to dress any wounds, however small, immediately, and not to eat, drink or smoke in the work zone.

The environment is best protected by containment of radioactive substances as in 5.3 below, and by its eventual disposal in authorised and regulated sites.

5.2.3. Subsequent actions to be taken

Reporting of alarm incidents is dealt with in 5.7 below.

A substantial metals recycling business with radiation detection equipment managed along the foregoing lines, will probably detect radioactively contaminated material. According to actual experience, the frequency might be less than one item per year. Where possible contaminated material will be returned to the originator, as suggested in 5.6.1 below. However, if not returned to the originator, over time some accumulation is to be expected. It is not best practice to leave such contaminated material untreated for long periods. In most countries with a nuclear industry, there will be one or more radwaste treatment facilities which have the equipment and expertise to collect such radioactive arisings, treat them appropriately and deliver them to an authorised disposal site. They also have the capability to check the storage area for any residual contamination, and to decontaminate if necessary. See also 5.5 below.

It would be prudent for any metals-recycling organisation to contact, in advance of any problems, their national or regional source of guidance and expertise on radioactivity and radiation protection.

5.3. CONTAINMENT OF THE RADIOACTIVELY CONTAMINATED MATERIALS

5.3.1. Dispersible and non-dispersible radioactive substances

With all hazardous materials, it is important to prevent the spread of the hazard. Dispersion can take place, for example, from spillage during transport, from liquids reaching ground water and sewerage systems, and from airborne dust and gases. Radioactive substances are no different. It is vital to prevent the release of radioactive material, and to contain any activity as soon as possible after it has been detected.

Radioactive material from manufactured radioactive sources is usually encapsulated (sealed) in a form that enables it to fulfil its function, such as for gauging, non-destructive testing, medical uses and the like. However, unsealed sources are also manufactured and used, especially for medical purposes. Both types of source are normally kept in a housing providing sufficient shielding to protect the user and the public. Such containers should normally be labelled. If the source is removed from its shielded container, then there is a hazard from external radiation exposure. However, if the source itself is ruptured resulting in the radioactive material becoming loose then there is also a contamination problem with its associated internal hazard. This is why visual recognition of the highly distinctive and prominently labelled packages is an important safeguard. It may sometimes be easier to pick out such packages with radioactive sources visually than to detect them with monitoring instruments. If they remain undetected and the shielding is destroyed during scrap processing, the consequences of the loss of control can be serious.

The concentration of NORM generally results from the processing of minerals, and is therefore nearly always found on the internal surfaces of process equipment as surface contamination (e.g. scale). The radiation dose from these deposits is usually low, but they are hazardous if inhaled or ingested, and they are susceptible to dispersion, for instance by rainwater washing the contamination out, or by the dust being blown around.

5.3.2. Protection packages (short-term and long-term)

Following detection of unwanted radioactive contamination, the simplest and most versatile containment will often be a thick plastic sheet wrapping sealed with tape. Larger items such as pipes with internal NORM contamination need only be sealed with plastic sheet and tape over their ends. For larger, heavier or sharper objects, a closed box of some kind may have to be utilised as containment. Rarely, the external dose rate may require shielding to be incorporated, using lead sheet or thick steel.

Such packages can only be considered as short-term expedients. In the longer term such problems as corrosion of metal containers and exposure to the weather make it necessary to dispose of the contaminated objects properly.

5.3.3. Transport packages

The safe transport of radioactive material, even low-level radioactive material, is ensured by compliance with the appropriate regulations, published by the IAEA.

Such regulations control the design of the packaging of radioactive material, and place limits on the external radiation levels. Annex 4 contains further detailed information.

The complexity of these regulations is usually beyond the competence of the recycling- and metal-producing industry and so specialist help will be required.

5.4. PERSONNEL PROTECTION

Recycling businesses of sufficient size may wish to include in the training of their staff some knowledge of elementary radiation protection, particularly the importance of the simple workplace rules (see 5.2. above).

In the event of a serious incident, expert help is essential, and enquiry to the competent governmental authorities of the country should lead to a source of specialist help.

Although it is most unlikely that any radioactivity detected in recycled material will give rise to the need for medical care, provisions have to be made to know where such medical care can be obtained.

5.5. DECONTAMINATION

Decontamination in the metals recycling context would have the objective of removing sufficient radioactive substances from a contaminated bulk of material to bring the bulk to a level of activity deemed to be acceptable by the regulatory authority, while leaving a much smaller and more contaminated residue for disposal at a regulated site. It will usually be carried out by a specialist contractor, whose advice will be valuable. The methods used rely on the contamination being on the surface, so that some form of surface cleaning is applied.

Recyclers and the metal works are not expert in decontamination and will therefore need to ask for expert help as the occasion arises.

5.5.1. Decontamination at the site of detection

Two scenarios present themselves, depending on whether containment has or has not been successful. Where the activity is contained, the option is decontamination on-site by a specialist contractor, or to transport the contaminated material as in 5.3.3 above to a radwaste treatment operation. Generally, small amounts are best tackled on-site, and larger tonnage quantities sent to a facility as in 5.5.2 below.

If radioactive substances have spread into a plant before detection, then on-site decontamination will be the only option, apart from disposal of plant items as low-level radwaste. Much will depend on individual circumstances, but if most contamination can be isolated to such replaceable items as conveyor belts, ladles, skips, furnace linings and filter media, then decontamination of the rest may be practical and economical.

5.5.2. Decontamination in a facility

In most nuclear-capable countries there are specialist decontamination facilities associated with their nuclear operations. Although primarily set up to decontaminate internal arisings, most do offer a commercial service, albeit at premium rates. The methods used still rely on removal of a surface layer by one or other means, and are therefore limited by the geometric complexity of the items to be processed. In every case, it will be prudent to compare the cost of decontamination with the cost of direct disposal. For example, NORM in large diameter straight pipes is easily decontaminated, but valves with the same contamination are probably best sent directly to a specialised disposal site.

5.5.3. Decontamination of soils and loose wastes

Although methods have been developed for partial decontamination of loose flowable wastes, particularly sandy soils, these are only economical on a fairly large scale, and in most cases the most cost-effective solution is direct disposal in a specialist facility.

5.5.4. Decontamination of personnel

If the precautions of 5.2. above are carefully followed, contamination should be limited to work clothing, overshoes, gloves and face masks. These will normally be collected, contained, and sent for disposal. Some nuclear facilities have contaminated clothing laundry facilities, but most protective clothing is treated as disposable.

In the event of contamination to the body, most commonly the hands and face will be affected. Careful frisking with an appropriate handheld probe should locate the contamination fairly precisely. It can then be swabbed off with one or more lanolin treated wipes, which are then disposed of as contaminated waste. If normal washing is needed, the washing water also becomes contaminated waste.

5.6. TRANSPORTATION (MATERIAL TRANSPORTED AFTER DETECTION, CHARACTERIZATION, AND CONTAINMENT OF THE RADIOACTIVE SUBSTANCES)

5.6.1. Conditions for the return of radioactively contaminated recyclable metal to the point of origin

By adhering to the national and international regulations that are applicable to the transport of radioactive materials (cf. annex 4) there is the possibility of safely returning it to its origin. However, as suggested in 5.1.3. above, this may mean that a larger organisation, that is in fact better able to detect and properly deal with the radioactivity, returns it to a smaller and less capable one. This could lead to the risk of illegal dumping. It may be in the best interests of the environment for countries to facilitate disposal of radioactive substances directly from where they are found, rather than have them escape control, perhaps later to surface again elsewhere.

5.6.2. Transfrontier shipment

The IAEA or similar regulations provide for safe transport, but do not address the export and import restrictions which different countries may apply. Many countries are reluctant to accept the radwaste of others, although there are some few countries prepared to import and treat low-level radwaste.

However, if the contaminated metal can clearly be shown to have originated in a particular country, and the contract allows for return to the point of origin, then re-export from the country of detection back to the country of origin is clearly equitable. It is desirable that no hindrance be placed by the re-importing country on such shipments once the safety standards for the transport have been met.

If such re-export involves transit through other countries, then the consignment should be treated as any other radioactive material shipment and no additional hindrance should be placed on such transports.

5.7. REPORTING

An initial report has the objective of informing the relevant authorities of an incident of a certain magnitude, so that they can begin to take any necessary action immediately or provide any requested advice.

A follow-up report would demonstrate that the operator has carried out the responsibilities laid down in 5.2.3 above. It could be used to follow up all incidents to ensure safe disposal of the radwaste found and to pursue the 'Polluter pays policy'.

As discussed in more detail in the next Chapter, all parties involved (the IAEA, the national regulatory bodies, and the metal recycling industry) have an interest in understanding the magnitude of the various problems associate with orphan sources, including their discovery, accidents arising from their discovery, and events involving the unintentional smelting of sources. It therefore, has a number of voluntary reporting mechanisms. However, the only obligatory reporting requirement is that under the Convention on Early Notification of a Nuclear Accident. This requires a State Party to notify the IAEA of an accident that may lead to an international transboundary impact of radiological significance for another State.

Two factors need to be considered before imposing any further reporting requirements on scrap metal operations, these are the aims of such reports, and the ability of the recipients of the reports to work effectively towards those aims. Aims, which would be widely supported by the industry include reducing the quantity of orphan radioactive metals in circulation, relief from operators bearing the costs of incidents, and protection of workers, the public and the environment.

5.7.1. Agencies to receive reports

Since nuclear regulations are dealt with country by country, the priority would be to report to the competent nuclear safety or nuclear regulatory authority of the country in which the incident occurs. If no such body exists, then a report to the authority dealing with environmental protection would be appropriate.

An event involving an international transboundary impact of radiological significance for another State should be reported by the designated authority to the IAEA Emergency Response Centre as required by the Early Notification Convention.

It would be desirable for a copy of the reports of found sources of categories I and II, to be provided by the national competent authority to the IAEA in order to complete the IAEA database on missing and found sources.

5.7.2. Reporting persons

The operator initiating the report would be the one with the detection system, the person responsible for the detection system being trained to make the report. The reporting person needs to know the required reporting format and its destination.

5.7.3. Elements

It is useful if the initial report contains the following elements:

- *Who is making the report?*
- *Where did the relevant batch of metals come from?*
- *What is the form of the batch?*
- *What quantity is involved?*
- *What is the mass of contaminated items?*
- *How and when was the radioactivity detected?*
- *What is the nature of the radioactivity?*
- *What is the dose rate and distance of measurement from the source? (essential)*
- *What is the surface activity or volumetric activity? (preferable)*
- *Which radionuclides have been identified? (preferable)*
- *What is the estimate of the total activity (preferable)*
- *Where is the activity currently contained?*

Subsequent reports could contain updates on the previous items and the following additional elements:

- *Was the contaminated material re-exported? If so, where to?*
- *What arrangements have been made for disposal in the country of import?*
- *Was the contaminated material disposed of in the country of detection?*

5.7.4. Reporting format

Reports should be made in accordance with the regulatory requirements. In the absence of any specific requirements, it is desirable that an initial report is made by e-mail or by fax as soon as reasonably practicable.

Similarly, subsequent reports can be made by e-mail or by fax within 30 days or after the contamination has been removed and disposed of by an authorised competent body.

CHAPTER VI

INTERNATIONAL COOPERATION ON ORPHAN SOURCE ISSUES

6.1. THE NEED FOR COOPERATION AND INFORMATION EXCHANGE

Co-operation and regular exchange of information, domestically and internationally, is important for preventing, detecting and responding to a loss of control over radioactive materials. Several current initiatives by the IAEA are aimed at dealing with the orphan source problem and strengthening the regulatory infrastructure of Member States.

Cooperation and exchange of information between neighbouring countries, or with countries that have been identified as possible originators of found radioactive materials, is encouraged. This initiative will foster bilateral, regional and international cooperation and communication concerning unauthorised movement of radioactive materials as well as creating opportunities to review the existing control mechanisms.

Measures that can help solve some of the problems include secure handling and physical protection of radioactive materials, the implementation of efficient accounting and control systems, as well as appropriate detection and response programs.

Assistance to States from international organisations is also being organised and will include establishing permanent contact points, sharing information and expertise, organising and participating in technical meetings and supporting national programmes. For this purpose, States should communicate information on cases of unauthorised movement or seizures of radioactive materials to the relevant international organisations in accordance with their national legislation and channels of communication. Regional workshops have also been organised by international organisations with a view to providing relevant information to the manufacturers and users of sources and related devices.

6.2. IAEA ACTION PLAN RELATED TO EVENTS INVOLVING ORPHAN SOURCES

The primary purpose of the IAEA action plan (approved in 1999, revised and re-approved in 2001), is to enable the Agency to develop and implement activities that will assist Member States in maintaining and, where necessary, improving the safety of radiation sources and the security of radioactive materials over their life cycle. Consideration is given to fostering a safety culture, including the development of effective regulatory infrastructures, and to the education and training and oversight of those responsible for radiation sources and radioactive materials. In particular, the training of persons that use radiation sources or radioactive materials will hopefully lead to the development of an increased sense of responsibility and safety culture so as to ensure that operations are undertaken safely and the sources and materials are kept secure.

The initiatives regarding the safety of radiation sources and the security of radioactive materials, including the problem of orphan sources, are grouped according to seven areas which provide a logical division of tasks being carried out by the IAEA:

- Regulatory Infrastructures
- Source Management and Control, including the Management of Disused Sources
- Categorisation of Sources
- Response to Abnormal Events
- Information Exchange
- Education and Training

- International Undertakings

Activities within the Action Plan that are relevant to the current document include the following initiatives related to abnormal events, i.e. events associated with orphan sources and unauthorised movement of radioactive material.

➢ Guidance on national strategies and programmes for the detection and location of orphan sources and their subsequent management is under development.
➢ Guidance on the prevention, detection and response to inadvertent movement and illicit trafficking of radioactive materials is being finalised. The detection document covers the selection and use of detection and monitoring equipment at borders.
➢ Efforts are being made to further develop national capabilities for dealing with radiological emergencies.
➢ The IAEA's existing capabilities for the provision of assistance in emergency situations are being strengthened.

Three databases are being developed as part of the Action Plan and other, existing databases also have some relevance.

6.2.1. Missing and Found Source Database

The IAEA Secretariat examined the usefulness and feasibility of such a database. Rapid exchange of information on missing or found very hazardous sources is primarily in the interest of organisations, that have to intervene in case of a radiological emergency and that have to be aware that such sources may be out of regulatory control. As the IAEA's emergency response system already has a well-defined set of contact points whose interest is in nuclear accidents and radiological emergencies, it was decided that such a database would be implemented under the arrangements of the IAEA's emergency response system.

The system includes a standardised reporting mechanism also for missing sources. This reporting mechanism is strongly linked to the 'Categorisation of Sources', which was developed by the IAEA Secretariat under the Action Plan on Orphan Sources[20]. According to this categorisation the reporting mechanism requests that all missing and found sources belonging to categories I and II except low dose rate (LDR) brachytherapy sources be reported to the IAEA. These sources will be put into the missing and found source database. The IAEA Secretariat will also maintain a close link with other organisational units in the secretariat that have similar databases (e.g. the illicit trafficking database) to ensure that information is always known to the relevant organisational units. Accordingly reports of found sources of these categories will also be dealt with. The Secretariat will follow up such reports to get all the information that is needed to successfully identify the reported sources.

To serve the objective of being a rapid exchange of information, this database is available on-line under the web pages of the IAEA's Emergency Response Centre. These web pages are secure web pages, meaning that only users with user-ID and passwords will be able to access this web-site. User-IDs and passwords are given to all nominated contact points for the IAEA's emergency response system[21].

6.2.2. The IAEA Radiation Events Database (RADEV)

The overall objectives of RADEV are to:

(a) disseminate information on radiation events and feedback lessons learned in order to prevent future accidents, or mitigate their consequences should they occur;

---

[20] IAEA-TECDOC-1191.

[21] "Emergency Notification and Assistance - Technical Operations Manual", EPR-ENATOM-2000, IAEA, Vienna (2000)

(b)   provide a tool to help Member States, the IAEA and other organisations to identify priorities in their radiation safety programme to facilitate the efficient allocation of resources.

In order to achieve these general objectives a centralised RADEV database is being established at IAEA headquarters in Vienna to:

(a)   provide a repository of information on accidents, near-misses and any other unusual events involving radiation sources, not directly involved in the production of nuclear power or its fuel cycle;

(b)   categorise events in a standardised manner to facilitate the search for events fitting particular profiles, the identification of causes and the lessons to be learned;

(c)   provide a means to analyse trends in radiation events;

(d)   provide summary descriptions of events that can be used directly as training material.

It should be noted that RADEV is designed to capture lessons learned from radiation events and is not meant to be a real-time on-line database.

More details about RADEV are given in annex 2.

6.2.3. International Catalogue of Sealed Radioactive Sources and Devices, including Transport Containers

The system of data on sealed sources is being materialised by the development of a catalogue containing specific information on industrially manufactured radiation sources and devices, including a visual guide to enable identification of the generic type of product based on outward appearance (e.g. shape, size, mass). This catalogue would assist in identification of specific models in order to allow safe handling of these items.

The development of the catalogue involves the collection of information available not only within the IAEA but from institutions such as source and equipment manufacturers, users and regulatory bodies in Member States.

The target group for use of this catalogue will be widely varied, and includes regulators, professionals dealing with orphan sources, law enforcement organisations, the metal recycling industry, the metal producing industry and waste management companies.

The catalogue will facilitate the identification of design specifications based on limited information obtained from "found" radioactive sources or devices to allow safe handling of these items.

More information about this international catalogue is given in annex 3.

6.2.4. Other Relevant Information Collection

There are approximately 70 countries participating in the Illicit Trafficking Database (ITDb) programme whereby discovered radioactive materials, including nuclear materials, are reported to the IAEA. The focus is on nuclear materials being deliberately trafficked, however, other radioactive materials that are discovered are also reported and included in the database.

Finally, while there is no formal reporting requirement or database, the IAEA has an interest in hearing about events involving the melting of radioactive sources, or the discovery of products made from raw materials in which a source has been melted. The IAEA is able to perform an interaction and co-ordination role when such information is provided and may be able to help minimise the consequences of such events by further distribution of the information. (Fax: +43 1 2600 7 29309; Phone: +43 1 2600 22025; Email: eru3@iaea.org)

CHAPTER VII

FINAL CONTROLS ON THE
PRODUCED METAL, SLAG AND OFF-GAS DUST

7.1. THE AIM OF RADIOACTIVITY CONTROL ON PRODUCTS OF THE PROCESS

When a radioactive source is melted in a furnace, the radioactive substances are not destroyed and the radioactivity remains. It becomes distributed between the metal, and the slag and the off-gas dust, which form in the process. The distribution of the radioactivity between these three phases depends on the chemical and physical properties of the radioisotope involved. For example in steel manufacture cobalt 60 is almost wholly absorbed by the steel, uranium passes to the slag, and caesium 137 passes to the off-gas dust. Figures 7.1, 7.2, 7.3 and 7.4 present examples of the relative part of radionuclides in metal, slag and exhaust gas, for the melt of stainless steel, for carbon steel and for a copper-based alloy.

Figure 7.1. [22]

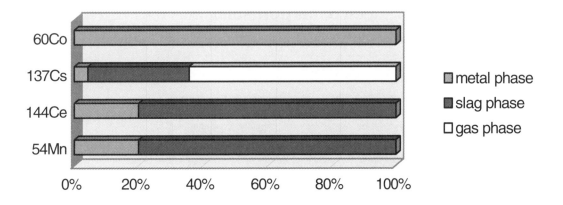

---

[22] Data from the Russian Federation.

Figure 7.2.[23]

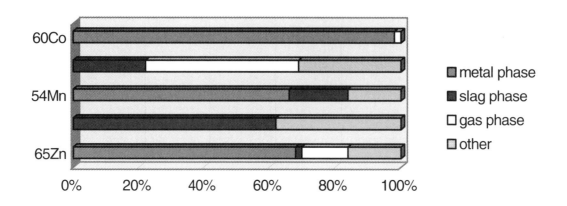

Figure 7.2.[23]

**Relative part of radionuclides in metal, slag and exhaust gas
Melt of carbon steel**

Figure 7.3.[24]

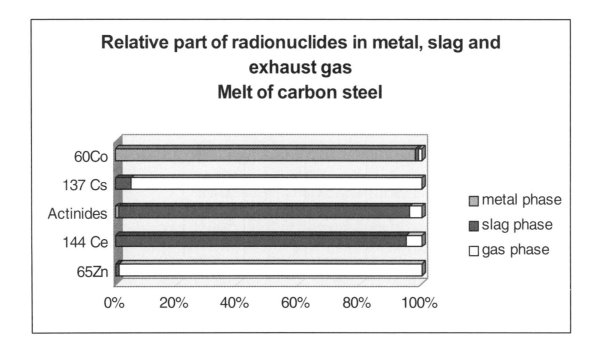

**Relative part of radionuclides in metal, slag and exhaust gas
Melt of carbon steel**

---

[23] Journal of the Randec n° 9, December 1993 – p. 46 – Nakamura & al.

[24] D.S. Harvey "Melting of contaminated steel scrap from decommissioning." Proceedings of a European Community conference 'Decommissioning of Nuclear Installations – Brussels – OCT. 1989.

Figure 7.4.[25]

Melt of Cu base alloy

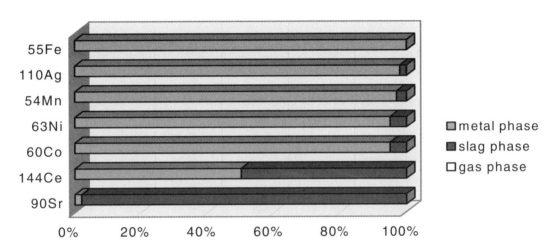

There have been a number of incidents in which cobalt 60 sources contained in steel scrap have been melted in steelmaking furnaces, and the resulting product, containing cobalt 60, has been distributed to customers. Because of these incidents, and the knowledge that the monitoring of the incoming scrap for radioactivity is still imperfect, it is beneficial to monitor the products of the process for radioactivity.

Specifically, there are three main purposes for monitoring the produced metal, slag and off-gas dust. The foremost of these is to ensure protection of workers and the public. In the event of an accidental source melting there would normally be a significant amount of dilution of the radioactivity; nonetheless it is still possible that radiation and/or contamination levels could be high enough to warrant careful control and to minimise personnel and/or public exposure.

The second purpose would be to measure for any radioactivity in the product, the by-products and the waste materials to determine what to do with them in the event of contamination. In most countries there is a legal definition of clearance levels of radioactive substances in materials. For materials that are traded internationally, the producer will have to supply material that satisfies the legal level in the country of the customer. In recent years this has resulted in some confusion over how to define a suitable level. Some national authorities and/or customers have requested a level defined in terms of radioactivity concentration (becquerel per gram). Others have requested levels defined in terms of dose rate at the surface of the material (microsievert/hour). There is broad agreement amongst radiation specialists that the level of radioactivity concentration (becquerel per gram), is more appropriate and should be adopted as the basis of the definition. Materials that are traded internationally should comply with a level of radioactivity which is acceptable by the trading countries. Hence they should comply with national laws, which should normally be based on the international safety standards developed by the IAEA (cf. § 3.3.).

The third reason for monitoring the product and by-products for radioactivity is to enable the producer to provide assurance to the customer that the products/by-products meet the specifications required by the customer. For example, the metalworks can state that the product is free from any measurable radioactivity in excess of the traces which are naturally present in all materials.

---

[25] Data from the Russian Federation.

7.2. DETECTION SYSTEMS FOR RADIOACTIVE CONTAMINATION OF PRODUCTS

7.2.1. Metal

It is routine in metal production to take samples of each melt for chemical analysis (in the steel industry, these samples typically weigh 70g). The same samples can be used to monitor the melt for gamma radiation.

There are established methods of measurement based on scintillation detectors (see section 4.2.3) that can achieve the necessary sensitivity of detection. Systems are commercially available from a number of manufacturers. A typical unit has a detector of sodium iodide (NaI) of 50mm diameter, and 50mm length. The detector must be sensitive over the range 50 keV to 1400 keV in order to be able to detect the most commonly encountered radionuclides. The detector is housed in a shield of lead weighing ~200kg. The shield serves to reduce the level of environmental background radiation, and so increase the sensitivity of detection. The detector is connected to instrumentation, which determines the level of radioactivity present. The measuring system will typically include software for background subtraction, alarm level setting with given accuracy, and an alarm when a high level is detected. The system is calibrated using very small radioactive sources of known activity. A system of this kind is able to measure the presence of cobalt 60 in steel at a level of at least 0.1 Bq/g within a few minutes. For other commonly encountered radionuclides the level of sensitivity is likely to be poorer since both the gamma energy and quantity of the emissions are likely to be less than those for cobalt 60.

Radionuclide identification is desirable since, unless the radionuclide can be identified, it is not possible to define the activity concentration (Bq/g). Some commonly encountered radionuclides, including cobalt 60 and caesium 137, can be identified with some confidence using instruments of the kind described above within a metallurgical laboratory.

It is, however, difficult under practical commercial conditions in a metal-producing plant to make precise measurements which cover the whole range of possible radionuclides that might be present in the sample. Hence it is advisable to set the alarm threshold corresponding to a level well below the legal levels and close to the typical natural background level in the metal. However it needs to be sufficiently above natural background to avoid frequent false alarms. Setting such a low alarm threshold is unlikely to cause any difficulty to the metal producer since metal products typically have a natural background radioactivity content below 20 - 50 Bq/kg, and will not cause false alarms. If an alarm occurs, the sample can then be investigated in more detail. This investigation might have to be done by a specialist laboratory using, for example, high-resolution gamma ray spectrometry.

While this investigation is in progress, it may be necessary for the metal-producing plant to put into action a plan to contain the radioactive substances, and to ensure that the exposure of people is minimised (cf. Chapter 5). This plan will have to be based on pessimistic assumptions of the likely hazard from the radioactive materials, since it is likely that the plan will have to be implemented before full information is available.

7.2.2. Slag

The system used for monitoring the level of radioactivity in the metal can also be used for monitoring the level of radioactivity in samples of slag. Slags are usually naturally slightly more radioactive than metal. The main practical difference is likely to be that the calibration for slag samples will be different to the calibration for metal samples.

In steel production, and most other metal-melting processes, the actinides and some other elements (^{54}Mn, ^{90}Sr, ^{144}Ce, ... – cf. section 7.1), are absorbed by the slag. Some of the actinides are emitters of alpha radiation, and emit very little gamma radiation. Rapid detection of alpha radiation in slag is very difficult and there is no proven method that can be routinely performed on-line at metal-producing plants. Monitoring for these elements using the gamma-detection equipment described will achieve a very poor sensitivity and low probability of detection.

Another easy way to detect additional radioactivity in slag is to use the fixed detectors at the entrance of the metal works that are used for the control of incoming scrap. Before leaving the metal work, trucks loaded with slag can pass before these detectors using the same procedure as for the control of the scrap, but with an appropriate alarm threshold (cf. section 4.5.).

7.2.3. Off gas dust

Some of the radionuclides like caesium 137 will pass mainly into the fumes of the melting shop and, after cooling of these fumes, into the off-gas dust.

Samples of off-gas dust can be examined using the same system that is used for metal. In practice, however, samples of off-gas dust are not routinely taken. The generally preferred method of monitoring for radiation in the off gas dust is therefore to have a detection system mounted on the gas-cleaning system. Detector systems of this kind are not yet in routine use at metal-producing plants. Those which are installed monitor the dust after it has been separated from the gas stream. At this point the dust is present in large amounts and a good sensitivity of detection can be achieved. An alarm can be given before the dust is removed from the gas cleaning system, so any radiation hazard can be minimised.

The principles of operation are similar to those of the detector systems used for scrap and for metal samples.

CHAPTER VIII

CONCLUSIONS AND RECOMMENDATIONS
FOR THE IMPROVEMENT OF THE SYSTEM

8.1. CONCLUSIONS

The issue of radioactive materials presents several problems for the scrap metal recycling industry and the metal producing industry. Principally, the introduction of either discrete sources, or of improperly released radioactively contaminated material, presents a health hazard for workers and a potential environmental hazard due to radioactive contamination of equipment, grounds, products and by-products. Subsequently, the health and safety of consumers of contaminated products may be put at risk.

Efforts have been made to protect the health and safety of workers, the environment and consumers. However, since incidents of radioactive contamination of metallurgical scrap continue to arise, further action is required to minimise risk.

The IAEA has published a series of documents, including the Basic Safety Standards[26], which includes guidance regarding the clearance of materials. It has also embarked upon a revised Action Plan for the Safety and Security of Radiation Sources. This IAEA programme includes a number of actions, such as an international database of missing and found radiation sources and the publishing of an international Code of Conduct on the Safety and Security of Radioactive Sources. Implementation of the Code of Conduct would likely prevent the introduction of radioactively contaminated materials into recycled metals. In addition, the IAEA is rapidly pursuing issues related to the clearance of materials, including those containing NORM in order to provide further international consensus and guidance.

In turn, the scrap recycling industry and metal producing industry have already invested considerably in the training of personnel and in the installation of detection equipment, thus providing among others a service and safeguard to society.

One of the goals of this document is to present recommendations to avoid the introduction of discrete sources and improperly released radioactively contaminated material into the recycling stream. Taking into account such recommendations, the probability of undesirable radioactive materials entering into the recycling loop will be minimised. This will ensure a better protection of the workers as well as of the public and the environment.

Another goal is to provide recommendations that allow the scrap recycling and metal production industry to make informed decisions on the purchase and use of the material that is properly released from specific activities such as the nuclear industry. This will assure the customer of the recycled metal product that the purchased product meets the desired specifications regarding radioactivity, thereby maintaining the consumer's confidence in the supplier.

---

[26] "International Basic Safety Standards for protection against ionising radiation and for the safety of radiation sources", published by the International Atomic Energy Agency under the Safety Series n° 115. This document is jointly sponsored by the Food and Agriculture Organisation of the United Nations, the International Atomic Energy Agency, the International Labour Organisation, the Nuclear Energy Agency of the Organisation for Economic Co-operation and Development, the Pan American Health Organisation and the World Health Organisation.

8.2.   RECOMMENDATIONS FOR IMPROVEMENT OF THE SYSTEM

Radioactive material may be introduced in metallurgical scrap through three different pathways.

> <u>Discrete radioactive sources</u> may be introduced into the scrap, due to the fact that such radioactive sources may escape from regulatory control because they are abandoned, lost, misplaced, stolen or otherwise transferred without proper authorisation.

> <u>Uncontrolled radioactively contaminated material</u> may appear in the scrap stream from the process where the material has been used. The material may have become contaminated after contact with either natural radionuclides or man-made radionuclides. One example of this might be in an extraction industry where scale containing naturally occurring radioactive material (NORM) that is deposited in pipes or equipment may not be under regulatory control in the first place. Another example could be improperly released material that has been used in the nuclear industry and was contaminated with man-made radionuclides above regulated limits.

> <u>Introduction of material with a very low level of radioactivity, released in accordance with the national regulatory framework.</u>

Three main topical areas related to the risk of introduction of radioactive materials into the scrap recycling process have been identified where improvements -should be made by international organisations, national Governments and industry, namely:

> <u>Prevention</u> of the introduction of radioactive materials.

> <u>Detection</u> of such an introduction (measurement, procedures and voluntary provision of information).

> <u>Reaction</u> capability to cope in the event of a detection of such an introduction.

These topical areas are developed hereafter for each of the three introduction pathways mentioned before.

8.2.1. Introduction of discrete radioactive sources

Discrete radioactive sources may be introduced into the scrap, due to the fact that such radioactive sources may escape from regulatory control because they are abandoned, lost, misplaced, stolen or otherwise transferred without proper authorisation. Such sources could cause both external or internal exposure as well as significant economic impacts. External exposure can occur due to physical contact or close proximity to the radioactive material. Internal exposure can occur due to direct contact with or processing of any uncontained material that can cause a worker to inhale, ingest or absorb the radioactive material. In addition, the introduction of such material may generate significant business disruption and financial loss due to detection and rejection of the material or from the handling and processing of undetected material which could contaminate equipment, grounds, products and by-products.

<u>Prevention</u>

The main measure to prevent entry of radioactive material into metal scrap is to ensure adequate control over the discrete sources by users and the national regulatory authorities.

The IAEA has developed standards of safety for protection against ionising radiation and radioactive sources. Their application significantly contributes to the prevention of radioactive sources from becoming orphaned. All Governments are strongly encouraged to expeditiously implement or strengthen their own regulations based on these standards, and to strictly enforce these regulations. In particular, the Code of Conduct for the Safety and Security of Radioactive Sources is recommended for implementation by all States.

<u>Detection</u>

Scrap recycling and metal producing companies, with the assistance of trade associations, national radiological protection organisations and suppliers of radiation sources, are encouraged to:

- train their personnel;
- develop procedures for visual inspection of scrap, principally during collection, in order to find discrete sources at their point of entry to, or early in the recycling industry; and,
- install and use detection equipment according to the manufacturer's specifications and to the recommendations of this document. Detection equipment should be installed at crucial points in the recycling loop, in particular, prior to locations where handling, processing or melting of the scrap could damage a source.

<u>Reaction</u>

Sealed sources may well have labels and markings indicating past ownership, or certification, as well as serial numbers that give the possibility of tracing an original user. However, it is in the nature of the demolition sector that even if an investigation were undertaken, the original owner may turn out to be a now-defunct business.

The issue of the management and disposal of orphaned sources that have been discovered needs to be addressed urgently by each State's regulatory body. The IAEA Code of Conduct makes the statement that "Every State should ensure that its regulatory body …is prepared, or has established provisions to recover orphan sources and to deal with radiological emergencies and has established appropriate response plans and measures". Some considerations in this respect include:

- guidelines for identifying and characterising such sources;
- arrangements for identifying appropriate destinations for managing their storage or disposal;
- adequate arrangements for transporting them to such destinations, including return of materials across national borders.

Competent authorities should make provisions to clearly allocate responsibilities associated with managing discovered orphaned sources. It is regarded as unfair to place the cost burden of storage and disposal of orphan sources, or clean-up of contamination caused by them, on the facility that finds them.

For new sources, the producer responsibility concept may also be used with the costs of the end-of-life management of sources internalised in their initial selling price. This removes the financial burden from the last owner or holder of source.

Several countries are providing a free-of-charge method of disposal for orphan sources as a means to encourage their detection and appropriate disposition. It is recommended that such a procedure be applied world-wide. In addition, efforts to require the return of sources to the supplier at the end of life are to be encouraged since this also decreases the probability of a source becoming out of regulatory control. The use of voluntary agreements and codes of practice as exemplified by the Spanish protocol (cf. annex 5) is recommended on a national and international basis

The scrap recycling and metal producing industry is encouraged to actively provide information to the reporting mechanisms set up by the IAEA following the discovery of orphaned sources. These data can then be analysed with a view to making further improvements and recommendations.

8.2.2. Introduction of uncontrolled radioactively contaminated material

The introduction of uncontrolled material into the recycling stream that is radioactively contaminated with either natural radionuclides or man-made radionuclides could pose similar health and economic impacts as those for discrete sources. This would also cause significant business disruption and financial loss due to detection and rejection of the material or handling and processing of material which could contaminate equipment, grounds, products and by-products. However, such impacts would usually be of a lesser magnitude than those encountered with discrete sources but at a greater frequency.

Prevention

The main measure to prevent entry of uncontrolled radioactively contaminated material into the metal scrap stream is to ensure adequate control over the materials coming from these specific activities or industries. This has to be supervised by the national regulatory authorities based on the recommendations of the IAEA, as set up in its standards and guidelines, especially the Basic Safety Standards.

A special case concerns industries whose activities deposit NORM at detectable levels onto materials that could be recycled. This can be the case for example in the extractive industries, notably those dealing with oil, gas, coal and phosphate. Since NORM is generally not regulated, NORM contaminated materials are often freely sold into the open market. This undesirable situation is being addressed by the IAEA, which is rapidly developing further recommendations regarding which materials should come under the scope of regulatory control. In the meantime, it is a practical desire of the scrap metal industry that they be informed by those selling or disposing of such materials whenever there is good reason to believe that due to origin or function, the particular materials are likely to be NORM contaminated. This avoids the problems associated with rejecting material after NORM contamination has been detected at a receiving operation, perhaps after passing through several hands. It is desirable that such contaminated materials be specifically identified and kept separate from the normal scrap recycling circuit so it does not enter unrestricted metal products.

Detection

Scrap recycling and metal producing companies, with the assistance of trade associations, national radiological protection organisations are encouraged to:

- train their personnel;
- develop adequate procedures, principally during collection, for determination of possible NORM contaminated materials based on their origin or function;
- install and use detection equipment according to the manufacturer's specifications and to the recommendations of this document. Detection equipment should be installed at crucial points of the recycling loop, in particular prior to locations where handling, processing or melting of the scrap may present an exposure potential to workers or the potential for contamination of equipment, grounds, products or by-products.

Reaction

The metal recycling industry is seriously disadvantaged with regard to these materials and requires assistance. Its operations are unfairly bearing a major share of the costs of detecting, characterising, segregating, storing and disposing of contaminated materials. It would seem appropriate that the concept of the "polluter pays" principle also be used for management of uncontrolled radioactively contaminated material and radioactive contamination caused by these materials. In this respect, issues related to the proper disposition of materials discovered to be contaminated needs to be treated in a similar manner to that of discrete sources discussed above.

Some considerations in this respect include:

- guidelines for identifying and characterising such material;
- arrangements for identifying appropriate destinations for managing their storage or disposal;

- adequate arrangements for transporting them to such destinations, including return of materials across national borders.

Competent authorities should make provisions to clearly allocate responsibilities associated with managing the contaminated material. It is regarded as unfair to place the cost burden of storage and disposal of these contaminated materials, or clean-up of contamination caused by them, on the facility that finds them.

The concept of the 'polluter pays' principle may also be used for the management of such contaminated material, alternatively a free-of-charge method of disposal for such contaminated material is recommended. The use of voluntary agreements and codes of practice as exemplified by the Spanish protocol (cf. annex 5) is recommended on a national and international basis.

8.2.3. Introduction of material with a very low level of radioactivity, released in accordance with a national regulatory framework

The introduction of low level radioactive material into the general recycling circuit, which is properly released according to the appropriate regulatory framework, is also of concern to the scrap recycling and to the metal producing industry. Because naturally occurring radioactive materials can cause alarm in the detection systems which might be considered from artificial origin, it is important to be able to trace the original supplier of the material to determine the appropriate response.

Prevention

Radioactive materials that are exempt from the requirements of the IAEA Basic Safety Standards or materials that are released from regulatory control do not have any significant radiological hazards associated with them. However, there is a perception that all radioactivity or all radiation is hazardous regardless of the level.

Therefore, as part of the contractual provisions and in order to satisfy the general customer demand, the metal recovery and recycling industry requires from the facility selling or disposing any metal with enhanced naturally occurring radioactivity or cleared from nuclear use, to be informed of this fact and the regulatory framework under which they have been released. Such information should be conveyed with the released materials to the successive suppliers and buyers of the metal scrap - up to and including the melting unit - to allow prior informed approval by the purchaser of the material.

Detection

Material released in accordance with appropriate national regulatory frameworks would have very low levels of radiation that are generally not detectable by commercial equipment used by the scrap recycling and metal producing industries. However, detection capabilities are continuously improving and may alarm with released material: therefore, it is advisable that all incoming scrap material be subjected to the same detection systems.

Reaction

Reactions following detection should be based on the contractual arrangements between seller and buyer.

8.2.4. Controls on the output materials from the metal works – metal, slag and off-gas dust

A final control on the output materials of the metal works, in particular the produced metal, the slag and the off-gas dust should be conducted, thereby providing additional assurance that radioactive materials have not been accidentally introduced into the plant.

It should be recognised that very sensitive detection equipment is needed for the final control of the produced metal because of the dilution that would have occurred when any unnoticed radioactive material was melted with a much larger bulk of clean material. Nevertheless, appropriate measurement of the produced metal will ensure that the final metal product meets the customer's specifications.

Activity: the number of nuclear transitions of a given amount of a radioactive substance in a unit of time. The unit of activity is the becquerel (Bq). One becquerel is equivalent to one transition per second. Formerly, activity was expressed in curie (Ci).

$$1 \text{ Ci} = 3.7 \; 10^{10} \text{ Bq}$$
$$1 \text{ Bq} = 2.7027 \; 10^{-11} \text{Ci}$$

Additional radioactivity in the metal: the radioactive substance content in the metal that is above the typical natural background radioactivity content.

Becquerel (Bq): the unit of radioactivity of a radionuclide. One becquerel is equivalent to one transition per second.

Collective dose: an expression for the total radiation dose incurred by a population, defined as the product of the number of individuals exposed to a source and their average radiation dose. The collective dose is expressed in man-sieverts (man.Sv).

Curie (Ci): a formerly used unit of radioactivity, now replaced by the becquerel (1 Ci = 3.7 10^{10} Bq).

Dose: generic expression related to the energy imparted by ionising radiation to the unit mass of organs or tissues. Radiation protection considers:

- Absorbed dose = energy absorbed per unit mass. The unit of absorbed dose is joule per kilogram (J/kg^{-1}) and its special name is gray (Gy).
- Equivalent dose = absorbed dose averaged over a tissue or organ, weighted by a factor related to the quality of the radiation.
- Effective dose = sum of the equivalent dose in all the tissues and organs of the body, weighted by factors related to the sensitivity of the organs or tissues. The unit of effective dose and of equivalent dose is also joule per kilogram (J/kg^{-1}) and its special name is sievert (Sv).

Dose rate: dose delivered in a specific unit time.

EU: European Union.

IAEA: International Atomic Energy Agency.

NORM: naturally occurring radioactive materials. Due to a particular process, these naturally occurring radioisotopes may concentrate in a particular product.

Orphan source: a source which poses sufficient radiological hazard to warrant regulatory control but is not under regulatory control, either because it has never been under regulatory control, or because it has been abandoned, lost, misplaced, stolen or transferred without proper authorisation.

"Polluter pays" principle: it is the principle by which the potential polluter must act to prevent pollution and those who cause pollution pay for remedying the consequences of that pollution.

Radiation: In this report, it means ionising radiation. For the purposes of radiation protection, radiation capable of producing ion pairs in biological material.

Radiation source means a radiation generator, or a radioactive source or other radioactive material outside the nuclear fuel cycles of research and power reactors.

Radioactive contamination: the presence of radioactive substances in or on a material or the human body where they are undesirable or could be harmful.

Radioactive source: radioactive substance capable of emitting ionising radiation, which is permanently sealed in a capsule or closely bonded and in a solid form, excluding material within the nuclear fuel cycles of research and power reactors. It also includes any radioactive material released if the source is leaking or broken.

Regulatory body: any body or bodies on which a State has conferred legal authority to regulate any aspect of the safety and security of radioactive sources, including legal authority to grant authorisations.

Regulatory control: any form of control applied to facilities or activities by a regulatory body for reasons related to radiation protection or to the safety and security of radioactive sources.

Sealed source: a radioactive source whose structure is such to prevent, under normal conditions of use, any dispersion of the radioactive substances in the environment.

Sievert (Sv): the special name of the unit of effective dose and of equivalent dose. One sievert is equivalent to one joule per kilogram.

UNECE: United Nations Economic Commission for Europe.

ANNEX 2

IAEA RADIATION EVENTS DATABASE (RADEV)

1. OBJECTIVES OF THE RADEV SYSTEM

The overall objectives of RADEV are to:

> ➢ disseminate information on radiation events and feedback lessons learned in order to prevent future accidents, or mitigate their consequences should they occur; and
> ➢ provide a tool to help Member States, the IAEA and other organisations to identify priorities in their radiation safety programme to facilitate the efficient allocation of resources.

In order to achieve these general objectives a centralised RADEV database is being established at IAEA headquarters in Vienna to:

> ➢ provide a repository of information on accidents, near-misses and any other unusual events involving radiation sources not directly involved in the production of nuclear power or its fuel cycle;
> ➢ categorise events in a standardised manner to facilitate the search for events fitting particular profiles, the identification of causes and the lessons to be learned;
> ➢ provide a means to analyse trends in radiation events;
> ➢ provide summary descriptions of events that can be used directly as training material

It should be noted that RADEV is designed to capture lessons learned from radiation events and is not meant to be a real-time on-line database

2. EVENTS TO BE INCLUDED

General Events

— events or potential events involving patients, workers or members of the public;
— events involving radiation sources which have been lost, found, stolen, or subject to unauthorised and inadvertent transfer/sale;
— events that occurred during the transportation of sources that resulted or could have resulted in the loss or degradation of control of radiation sources.

Events Involving Patients

Many types of radiation events involving patients have been reported, including:

— Wrong patient exposed
— Wrong tissue exposed (correct patient)
— Wrong radio-pharmaceutical administered
— Wrong activity administered
— Wrong beam settings
— Delivered dose different from intended

The consequences of such events include: ineffective treatment, ineffective diagnosis, severe radiation burns, severe degradation in quality of life and, in some cases death directly attributable to high radiation exposure. Many of these events were caused by deficiencies in, or a lack of: design, testing and calibration of equipment; education, training and qualification of personnel; procedures; defence in depth; quality assurance. In some cases, events involving patients have also resulted in exposures to hospital workers, lost sources and exposures to members of the public.

3. MANAGEMENT AND OPERATION OF THE SYSTEM

The database has been designed to operate on a personal computer using Microsoft Access 97 or above. Copies of the RADEV software will be provided to selected organisations within Member States for their own use and they will be requested to provide data to the IAEA on a regular basis. The IAEA will manage and operate the international RADEV database, which will act as central focal point for all users. The IAEA will publish regular summary reports from RADEV and will provide electronic updates of the data to participating organisations. Confidentiality will be maintained by the IAEA at all times and details such as names of individuals, hospitals and factories will not be divulged.

4. IMPLEMENTATION

The RADEV project is being implemented in three phases:

Phase 1: Collection of currently available details of radiation accidents followed by in-house testing of the software.

Phase 2: Limited international trials – the IAEA will provide a working version of RADEV to several international and national organisations (including professional organisations in the medical field) for testing and evaluation. Feedback from the trials will be reviewed by the IAEA and any necessary changes made to the software.

Phase 3: Distribution of RADEV. The IAEA will collect data from participating organisations, compile international statistics and produce summary reports. Electronic copies of the summary reports and the updated database will be available to participating organisations.

The current status is that Phase 1 has been successfully completed and international trials are taking place.

ANNEX 3

DESCRIPTION OF
WHAT RADIOACTIVE SOURCES MIGHT LOOK LIKE

INTERNATIONAL CATALOGUE OF SEALED RADIOACTIVE SOURCES AND DEVICES, INCLUDING TRANSPORT CONTAINERS

A system of data on sealed sources is being materialised by the development of a catalogue containing specific information on industrially manufactured radiation sources and devices, including a visual guide to enable identification of the generic type of product based on outward appearance (e.g. shape, size, weight). This catalogue would assist in the identification of specific models in order to allow safe handling of these items.

The development of the catalogue involves the collection of information available not only within the IAEA but from institutions such as source and equipment manufacturers, users and regulatory bodies in Member States.

The target group for use of this catalogue will be very varied, and includes regulators, professionals dealing with orphan sources, law enforcement organisations, the metal recycling industry, the metal producing industry and waste management companies.

This catalogue will facilitate the identification of design specifications based on limited information obtained from "found" radioactive sources or devices, to allow safe handling of these items.

The Catalogue tries to track a source if some of these data are known:

1) Radionuclide(s)
2) Source model number
3) Source design (i.e. shape, width/diameter, thickness/height, length, chemical form, capsule material)
4) Manufacturing details or feature description, including shipping methods
5) Manufacturer
6) Distributor
7) Source type (the application, e.g. radiography, etc.)
8) Associated devices (the devices known to contain this source)
9) Maximum activity allowed in the source
10) Beginning of manufacturing period for model
11) End of manufacturing period for model (if production discontinued)
12) Countries or geographical areas in which this model has been distributed
13) Categories of users (mention the characteristics of users, e.g. medical institutions, research laboratories, industrial sites, etc.).

Some of the following additional data are useful when sources are a part of a device, including transport containers:

14) Device model number
15) Device design (i.e. shape, width/diameter, thickness/height, length, weight, materials)
16) Manufacturing details or feature description, including shipping methods
17) Manufacturer of the device
18) Distributor of the device

19) Device type (the application, e.g. radiography, etc.)
20) Associated sources (the sources known to be contained in this device)
21) Maximum activity allowed in the device
22) Beginning of manufacturing period for device model
23) End of manufacturing period for device model (if production discontinued)
24) Countries or geographical areas in which this device model has been distributed
25) Categories of users (mention the characteristics of users, e.g. medical institutions, research laboratories, industrial sites, etc.).

Sometimes there can be multiple entities involved in the manufacture and distribution of a single product, starting with the source material producer, continuing through multiple steps of manufacture of the source as well as the device, and possibly several levels in a distribution chain before it reaches the first end-user. For these cases, it can be difficult to determine which manufacturer and distributor to list in the database. For the purposes of this Catalogue, the distributor should be the initial distributor of the product in question, and the manufacturer should be the entity involved in the final stage of manufacture of the product. Either the manufacturer or distributor listed should be the entity that maintains the information related to the product.

The above listed data are shared in fields of three tables in a MS Access system:

- Sources: table containing data on items 1) to 13).
- Devices: file containing data on items 1), 2), 14) to 25).
- Manufacturers: file containing addresses of companies of items 5), 6), 17) and 18).

Some particular information, not allocable into one specific field, is kept as "comments".

Other complementary tables, such as Units (Conversion factors), Years (Manufacturing or distribution periods of time), Countries (Names of countries or geographical areas), References, etc., help the Catalogue to track a radioactive source.

To proceed to the tracking of the source, the enquirer fills in a form displayed on the screen by selecting in each box the required item. These items are the characteristics of the tracked source. If the required item is not found in the list of a box or is unknown, the box is kept blank. When the choice is done in all boxes, the Catalogue will deliver the identity or the most likely identity of the source tracked. Several results can be obtained if the selection matches or approximates several identities.

The following are samples of different screens of the catalogue:

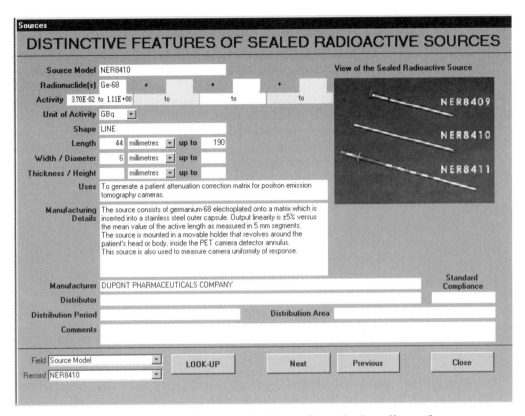

Record of the table on characteristics of sealed radioactive sources

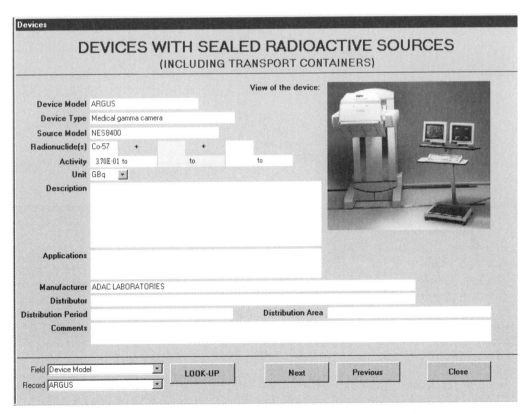

Record of the table on devices with sealed radioactive sources

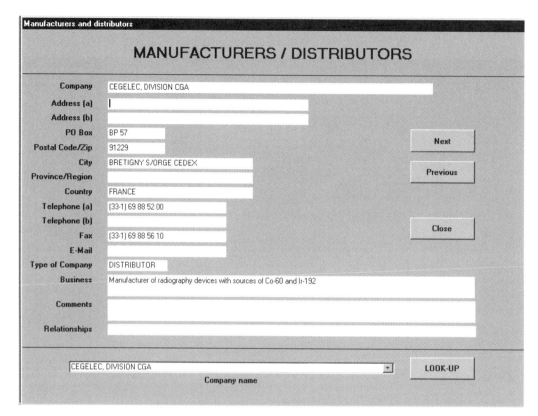

Record of the table with data on manufacturers and distributors

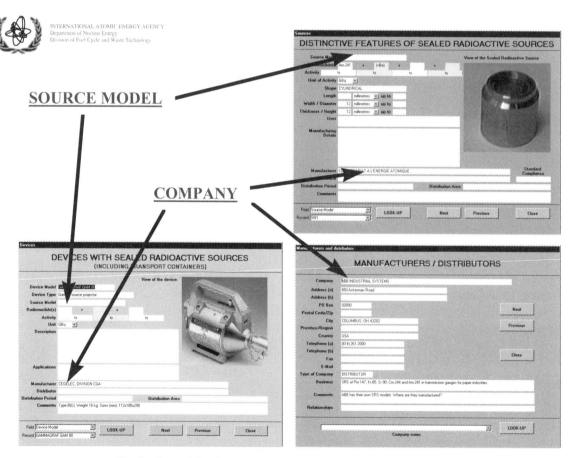

Relationship between the three main tables

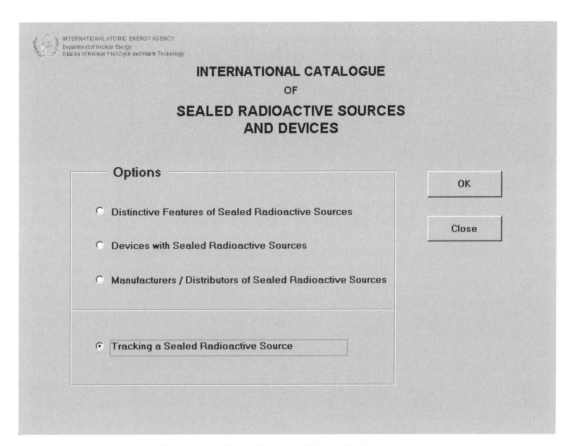

Screen of options of the Catalogue

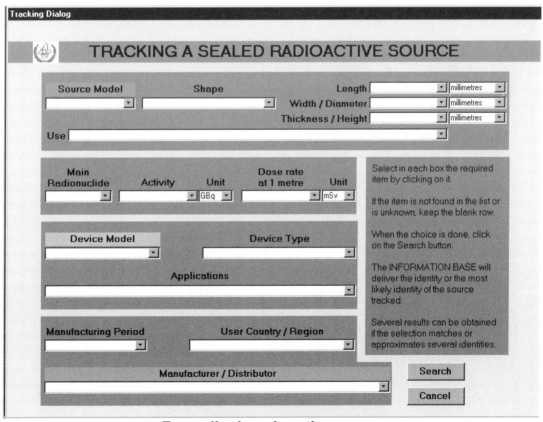

Form displayed on the screen

ANNEX 4

IAEA REGULATIONS FOR THE
SAFE TRANSPORT OF RADIOACTIVE MATERIAL

The following indicates the broad scope of the Transport Regulations as they should be applied to the transport of radioactive materials likely to be found in recycled metal. This should not be used as a substitute for complete application of the Regulations in a particular case. Specialist advice and consultation with relevant competent authorities will be needed to ensure adherence to the Regulations when recycled metals containing radioactive material are transported.

BACKGROUND

The IAEA Regulations for the Safe Transport of Radioactive Material, IAEA Safety Standards Series No. ST-1[27] establish standards of safety that provide an acceptable level of control of the containment, radiation, criticality and thermal hazards to persons, property and the environment that are associated with the transport of radioactive material. The IAEA Regulations serve as the model used by individual countries in developing their dangerous goods transport regulations, and by the international modal organisations such as the International Maritime Organisation (IMO) and the International Civil Aviation Organisation (ICAO) in developing their regulatory documents used in regulating the transport of all dangerous goods by sea and air, respectively, throughout the world.

The IAEA Regulations utilise the principles set forth in both the "Radiation Protection and the Safety of Radiation Sources", IAEA Safety Series No. 120 and the "International Basic Safety Standards for Protection against Ionising Radiation and for the Safety of Radiation Sources", IAEA Safety Series No. 115, jointly sponsored by the Food and Agriculture Organisation of the United Nations, the IAEA, the International Labour Organisation, the Nuclear Energy Agency of the Organisation for Economic Co-operation and Development, the Pan American Health Organisation and the World Health Organisation. Thus, compliance with the Transport Regulations is deemed to satisfy the principles of the Basic Safety Standards in respect of transport.

The Safety Standard ST-1 is supplemented by a hierarchy of Safety Guides and Safety Practices including "Advisory Material for the IAEA Regulations for the Safe Transport of Radioactive Material", IAEA Safety Standards Series No. TS-G-1.1 (ST-2) and "Emergency Response Planning and Preparedness for Transport Accidents Involving Radioactive Material", IAEA Safety Series No. TS-G-1.2 (ST-3) both expected to be published in early 2002; "Compliance Assurance for the Safe Transport of Radioactive Material", IAEA Safety Series No. 112; and "Quality Assurance for the Safe Transport of Radioactive Material", IAEA Safety Series No. 113.

In certain parts of the Regulations, particular actions are prescribed, but the responsibility for carrying them out is not specifically assigned to any particular legal person. Such responsibilities may vary according to the laws and customs of different countries and the international conventions into which these countries have entered. For the purpose of the Agency's Regulations, it is not necessary to make these assignments, but only to identify the actions themselves; it remains the prerogative of each Government to assign these responsibilities.

---

[27] Published in 1996. Updated in English in 2000, with minor corrections as TS-R-1 (ST-1, Revised).

OBJECTIVE

The objective of the Regulations is to protect persons, property and the environment from the effects of radiation during the transport of radioactive material.
This protection is achieved by requiring:

(a) containment of the radioactive contents;
(b) control of external radiation levels;
(c) prevention of criticality;
(d) prevention of damage caused by heat.

These requirements are satisfied firstly by applying a graded approach to content limits for packages and conveyances and to performance standards applied to package designs depending upon the potential hazard of the radioactive contents. Secondly, they are satisfied by imposing requirements on the design and operation of packages and on the maintenance of packaging, including a consideration of the nature of the radioactive contents. Finally, they are satisfied by requiring administrative controls including, where appropriate, approval by competent authorities.

In the transport of radioactive material the safety of persons, who are either members of the public or workers, is ensured when the requirements specified in the Regulations are complied with. Confidence in this regard is achieved through quality assurance and compliance assurance programmes.

SCOPE

The Regulations apply to the transport of radioactive material by all modes on land, water or in the air. Transport comprises all operations and conditions associated with and involved in the movement of radioactive material; these include the design of packages, manufacture, maintenance and repair of packagings, and the preparation, consigning, loading, carriage including in-transit storage, unloading and receipt at the final destination of consignments of radioactive material and packages. A graded approach is applied to the performance standards in these Regulations that is characterised by three general severity levels:

(a) routine conditions of transport (incident free);
(b) normal conditions of transport (minor mishaps);
(c) accident conditions of transport.

The Regulations do not apply to:

(a) radioactive material that is an integral part of the means of transport;
(b) radioactive material moved within an establishment which is subject to appropriate safety regulations in force in the establishment and where the movement does not involve public roads or railways;
(c) radioactive material implanted or incorporated into a person or live animal for diagnosis or treatment;
(d) radioactive material in consumer products which have received regulatory approval, following their sale to the end user;
(e) natural material and ores containing naturally occurring radionuclides which are not intended to be processed for use of these radionuclides provided the activity concentration of the material does not exceed 10 times the exemption levels specified in §§ 401–406 of Transport Regulations (same as in the IAEA Basic Safety Standards).

In addition, radioactive material is defined, for purposes of control through the Regulations, as any material containing radionuclides where both the activity concentration and the total activity in the consignment exceed radionuclide specific exemption values. These exemption limits are specified in §§ 401–406.

The Regulations do not specify controls such as routing or physical protection, which may be instituted for reasons other than radiological safety. Any such controls shall take into account radiological and non-radiological hazards, and shall not detract from the standards of safety, which these Regulations are intended to provide.

For radioactive material having subsidiary risks, and for transport of radioactive material with other dangerous goods, the relevant transport regulations for dangerous goods of each of the countries through or into which the material is to be transported shall apply in addition to these Regulations.

GENERAL PROVISIONS

RADIATION PROTECTION

The Regulations require a number of actions regarding radiation protection. A Radiation Protection Programme shall be established for any transport of radioactive material. The nature and extent of the measures to be employed in the programme shall be related to the magnitude and likelihood of radiation exposures. Programme documents shall be available, on request, for inspection by the relevant competent authority.

In transport, protection and safety shall be optimised in order that the magnitude of individual doses, the number of persons exposed, and the likelihood of incurring exposure shall be kept as low as reasonably achievable, economic and social factors being taken into account. In addition doses to persons shall be below the relevant dose limits specified in the Regulations. A structured and systematic approach shall be adopted and shall include consideration of the interfaces between transport and other activities.

The Regulations require that workers receive appropriate training concerning the radiation hazards involved and the precautions to be observed in order to ensure restriction of their exposure and that of other persons who might be affected by their actions.

The relevant competent authority shall arrange for periodic assessments of the radiation doses to persons due to the transport of radioactive material, to ensure that the system of protection and safety complies with the IAEA Basic Safety Standards (SS115).

For occupational exposures arising from transport activities, where it is assessed that the effective dose:

(a)   is most unlikely to exceed 1 mSv in a year, neither special work patterns nor detailed monitoring nor dose assessment programmes nor individual record keeping shall be required;
(b)   is likely to be between 1 and 6 mSv in a year, a dose assessment programme via workplace monitoring or individual monitoring shall be conducted;
(c)   is likely to exceed 6 mSv in a year, individual monitoring shall be conducted.

When individual monitoring or workplace monitoring is conducted, appropriate records shall be kept.

Radioactive material shall be segregated sufficiently from workers and from members of the public. The following values for dose shall be used for the purpose of calculating segregation distances or radiation levels:

(a)   for workers in regularly occupied working areas a dose of 5 mSv in a year;
(b)   for members of the public, in areas where the public has regular access, a dose of 1 mSv in a year to the critical group.

EMERGENCY RESPONSE

In the event of accidents or incidents during the transport of radioactive material, the Regulations require that emergency provisions, as established by relevant national and/or international organisations, shall be observed to protect persons, property and the environment. Emergency procedures are required to take into account the formation of other dangerous substances that may result from the reaction between the contents of a consignment and the environment in the event of an accident.

QUALITY ASSURANCE

The Regulations further require that quality assurance programmes based on international, national or other standards acceptable to the competent authority be established and implemented for the design, manufacture, testing, documentation, use, maintenance and inspection of all special form radioactive material, low dispersible radioactive material and packages and for transport and in-transit storage operations to ensure compliance with the relevant provisions of the Regulations. Certification that the design specification has been fully implemented shall be available to the competent authority. The manufacturer, consignor or user shall be prepared to provide facilities for competent authority inspection during manufacture and use and to demonstrate to any cognisant competent authority that:

(a)   the manufacturing methods and materials used are in accordance with the approved *design* specifications;

(b)   all *packagings* are periodically inspected and, as necessary, repaired and maintained in good condition so that they continue to comply with all relevant requirements and specifications, even after repeated use.

Where competent authority approval is required, such approval shall take into account and be contingent upon the adequacy of the quality assurance programme.

COMPLIANCE ASSURANCE

The Regulations also require that the relevant competent authority(ies) is(are) responsible for assuring compliance with the requirements of the Regulations. Means to discharge this responsibility include the establishment and execution of a programme for monitoring the design, manufacture, testing, inspection and maintenance of packaging, special form radioactive material and low dispersible radioactive material, and the preparation, documentation, handling and stowage of packages by consignors and carriers, to provide evidence that the provisions of the Regulations are being met in practice.

SPECIAL ARRANGEMENT

Consignments for which conformity with the other provisions of the Regulations is impracticable shall not be transported except under special arrangement. Provided the competent authority is satisfied that conformity with the other provisions of the Regulations is impracticable and that the requisite standards of safety established by the Regulations have been demonstrated through means alternative to the other provisions, the competent authority may approve special arrangement transport operations for single or a planned series of multiple consignments. The overall level of safety in transport shall be at least equivalent to that which would be provided if all the applicable requirements had been met. For international consignments of this type, multilateral approval shall be required.

TECHNICAL PROVISIONS

General package design requirements

The Regulations specify, in detail, general requirements for the design of all packages used in the transport of radioactive material. These can be broadly summarised as follows:

- Easily, safely transportable, designed for proper securing in or on vehicle
- Any lifting attachments properly designed with appropriate safety factors
- If these should fail, package would still meet Regulations
- Free from protrusions, easy to decontaminate
- Prevents collection and retention of water
- Not affected by acceleration or vibration
- Packaging compatible with contents
- Takes account of ambient conditions
- Takes account of any other dangerous properties of contents

Excepted packages

These usually contain very small quantities of radioactivity - e.g. instruments with radium painted indications. For each contaminant on an article or item, or in a material, the maximum quantities allowed in excepted packages are specified in the Regulations. For example, for a typical contaminant, Co(60), the maximum quantity inside an excepted package if the contaminant is not incorporated into an instrument or article is 400 MBq (11 nCi).

Excepted packages must meet the general packaging design requirements above, and also the following operational requirements:

- External marking UN 2910 (2911 for instruments, or 2909 for articles manufactured from natural uranium or depleted uranium or natural thorium), also gross mass if the gross mass exceeds 50 kg
- Internal marking to warn of radioactivity on opening
- Packaging which retains the radioactive contents during transport
- Non-fixed external contamination < 4 Bq/cm^2 (24,000 dpm/100 cm^2)(ten times less for alpha emitting nuclides)
- Surface radiation level < 5 microSv/hr (0.5 millirem/hr)
- Controls for damaged or leaking packages

Industrial packages

The next degree of complexity are Industrial packages (IPs), there are three types depending upon the type of material transported. The package must meet the general packaging design requirements above, and also the following operational requirements:

- External marking with «trefoil» international symbol for radiation
- Other detailed marking requirements regarding contents, etc.
- Must have smallest overall external dimension > 10 cm (4inches)
- Non-fixed external contamination < 4 Bq/cm² (24,000 dpm/100 cm^2)(ten times less for alpha emitting nuclides)
- Surface radiation level < 100 microSv/hr (10 millirem/hr) (more in some circumstances)
- Limits on specific activity if contents qualified as low specific activity material (LSA)
- Limits on contamination if the contents are qualified as surface contaminated objects (SCO)
- Limits on radiation level from unshielded material or objects and total activity in the conveyance
- Checks for contamination of conveyance
- Regulations concerning segregation
- Regulations concerning passengers

Type A, B and C packages

When the activity exceeds the limits for excepted packages or the contents cannot be qualified as either low specific activity material or surface contaminated objects for transport in industrial packages, Type A, B or C packages, with more stringent design requirements, must be used as appropriate. Type B and C packages require competent authority approval for design.

> Type A packages are designed to withstand normal conditions of transport; their radioactive content is limited in a generic way (the so-called A_1 and A_2 values).
> Type B packages are designed to withstand accident conditions of transport; their radioactive content is higher than A_1 and A_2. Their design requires competent authority approval; the approval certificate specifies limits on the content.
> Type C packages are designed to withstand accident conditions of air transport. Their design requires competent authority approval; the approval certificate specifies limits on the content.

Shipments of very large quantities of radioactive material, of large quantities of fissile material, or of special arrangements require prior competent authority approval.

SPANISH PROTOCOL FOR COLLABORATION ON THE RADIATION MONITORING OF METALLIC MATERIALS

PART 1. INTRODUCTION

THE MINISTRY OF INDUSTRY AND ENERGY (MINER)

THE MINISTRY OF DEVELOPMENT

THE NUCLEAR SAFETY COUNCIL (CSN)

LA EMPRESA NACIONAL DE RESIDUOS RADIOACTIVOS S.A. (ENRESA)

LA UNIÓN DE EMPRESAS SIDERÚRGICAS (UNESID)

LA FEDERACIÓN ESPAÑOLA DE RECUPERACIÓN (FER)

MINDFUL THAT:

- In recent years radioactive material has quite frequently been found in scrap, thereby giving rise to growing social concern in view of the risks involved.

- The recycling of metals to produce alloys of different compositions is an industrial activity that is extremely important for the economy and the environment.

- It is necessary to adopt radiation monitoring measures in an attempt to prevent and, where necessary, detect and control the presence of radioactive material in the scrap that some metallurgical plants use for the raw material in their production process.

- The implementation of monitoring measures in an attempt to exclude radioactive material from the metal materials used by metallurgical plants provides a mechanism for certifying that the product is free of radioactive contamination and, consequently, represents an added guarantee as regards the quality of the product in radiation terms.

- The primary object of the activities involved in the monitoring and control of the presence of radioactive material in scrap should be to prevent the inclusion of radioactive material in the scrap recycling process and, as a minimum requirement, to detect its presence as closely as possible to the point of inclusion.

WHEREAS:

- There is a variety of laws and regulations in Spain specifically regulating industrial activities involving nuclear and radioactive materials, and the possession, use and transfer of radioactive sources is regulated by the Nuclear Energy Act 25/1964, Act 14/1999 on Public Rates and Prices for services provided by the Nuclear Safety Council, and the regulations on Nuclear and Radioactive Facilities, approved by decree 2869/1972.

- This regulatory framework is unable to prevent deliberate or unintentional acts that result in radioactive materials being included in scrap.

- In view of the markedly transnational nature of the scrap market in our country, and the leading role that maritime imports play in this market, there is a need to put in place mechanisms to control metal products that enter the country through our ports.

- This issue is being considered by a number of international agencies that are studying the many different aspects of the problem. Since this is also a major concern for other Member States of the European Union, it would be appropriate for these States to agree upon the joint adoption of measures that they consider necessary to improve the control of the presence of radioactive materials in scrap. To this end, the Spanish Government has approached the European Commission with the request that it should promote the adoption of such measures, which, if appropriate, would have to be observed in the future.

- In the meantime, it is appropriate to establish a framework for action that determines the conditions in which the aforesaid measures should be implemented.

- Based on the results of the implementation of this Protocol or the development of international initiatives in this matter, essentially community initiatives, these monitoring measures could acquire legislative status in the future.

AGREE:

One. -   To sign this Protocol for Collaboration on the Radiation Monitoring of metal materials and final products defined in the Technical Annex, which is an integral part hereof, with a view to introducing the monitoring and control measures stated herein.

Two. -   To set up at the Ministry of Industry and Energy, for the implementation of this Protocol, a Register in which companies carrying out the activities referred to in the Technical Annex can register, thereby accepting the rights and obligations arising from registration.

Three.-   To foster the registration of companies in the Register referred to in the foregoing point, particularly the registration of companies that have facilities for the smelting or the storage and preparation of scrap.

Four.-   To consult every six months to analyse the results of the implementation or this Protocol and study possible amendments to the Technical Annex proposed as a result of this implementation.

Five. -   To appoint the Ministry of Industry and Energy as the depository of this Protocol, which will be kept open for accession by other industry associations involved in similar activities.

SIGNED BY ALL THE PARTIES CONCERNED

PART 2. TECHNICAL INFORMATION

1. Object

The object of this protocol is to establish requirements for the radiation monitoring of metal materials and final products, as defined in point 2, with the aim of detecting the presence of radioactive materials and of avoiding the risk of their dispersion and the consequent exposure to radiation or contamination of persons, property and the environment.

2. Definitions

For the purposes of the implementation of this Protocol, the following terms shall have the meanings defined below:

Signatory company

- The natural or legal person that carries out the activities referred to in point 3 and accedes to the Protocol.

Monitoring and control system

- The set of human resources and technical, organisational, operational, logistical and training facilities set up by the signatory company to detect and, where appropriate, separate and analyse radioactive materials that may be found in metal materials and final products, and to adopt such urgent measures as are necessary to avoid the dispersion of the radioactive material.

Metal material

- The scrap, ingots and metal semi-finished products that will serve as raw material for processing in facilities to which the Protocol applies.

Final product

- The products, semi-finished products, by-products and waste materials generated in the processing of the metal materials.

Radiological protection specialist

- A technician with certified knowledge of radiological protection and instrumentation who is employed by the signatory company or by a duly authorised Radiological Protection Technical Unit (UTPR).

3. Field of application

The Protocol is applicable to the following activities:

   a)   The recovery, storage or handling of metal materials for recycling,
   b)   The processing of metal materials.

4. Register of facilities covered by the Protocol.

The Ministry of Industry and Energy (MINER) will set up a Register of Facilities of the companies that have signed the Protocol.

The signatory companies will enter each of their facilities in the aforesaid register, and provide a declaration containing the information listed in annex 1.

The entry will be updated every five years and whenever a substantial change is made in the monitoring and control system or a change of ownership of the company occurs.

No charge shall be made for the entry in the Register of Facilities of the Ministry of Industry and Energy.

5. Undertakings arising from the implementation of the Protocol

5.1 The Ministry of Industry and Energy undertakes to:

a) Issue a standing order for the Authorisation of the Transfer to ENRESA of radioactive material detected in the facilities, in accordance with the legal provisions in force, subject to a report from the Nuclear Safety Council.

b) Establish and keep up to date the Register of Facilities of the signatory companies or advise the Nuclear Safety Council of the registration of each facility and notify these signatory companies of the registration of their facilities.

c) Carry out actions that are necessary to resolve situations that call for exceptional measures arising from the presence of radioactive material in metal materials and final products. Such actions will be adopted in coordination, where appropriate, with the other competent public agencies and companies concerned, subject to a report from the Nuclear Safety Council that will be mandatory and binding in matters falling under its authority.

5.2 The Ministry of Developments undertakes to:

a) Require the presentation of the certificate referred to in 5.5.b) as a requirement for the authorisation of the docking of the vessel.

b) Inform the Nuclear Safety Council of any incident of a radiological nature that occurs within its areas of authority when the transport of metal materials is involved.

5.3 The Nuclear Safety Council (CSN) undertakes to:

a) Advise ENRESA and the signatory companies of the application of the Transfer Authorisation where this is issued.

b) Issue such standing technical instructions and recommendations as it considers necessary for the implementation of this Protocol.

c) Take cognisance of the entries of facilities in the Register of the Ministry of Industry and Energy and, where appropriate, issue such technical recommendations or instructions as it considers necessary to guarantee that the reported monitoring and control system meets the requirements agreed in the Protocol.

d) Inspect the monitoring and control system set up by the signatory company and forward to the latter such instructions as it considers relevant to compliance with the requirements agreed in the Protocol.

e) Advise the competent authorities and the signatory companies on matters of radiological safety and protection with a view to compliance with this Protocol.

f) Promote the organisation of activity-focused radiological protection training and information campaigns among the employees of companies in the metal recovery and smelting sector.

5.4 The Empresa Nacional de Residuos Radioactivos (ENRESA) undertakes to:

a) Remove and hold the radioactive materials transferred to it after being detected in the facilities of signatory companies.

b) Provide technical advice to the signatory companies and collaborate with the latter in returning radioactive materials to the shipper where the latter is a foreigner.

c) Collaborate on training programmes for technicians who will be required to take action in the event of radioactive material being detected.

d) Collaborate on radiological protection training and information campaigns among the employees of companies in the metal recovery and smelting sector.

e) Sign a contract with the signatory company for the handling of radioactive materials in accordance with the provisions of point 6.3.a).

5.5 The signatory company undertakes to:

a) Carry out radiation monitoring of metal materials and final products. To this end:

- It will install, operate and maintain a system for monitoring and controlling radioactive material in metal materials and final products.
- Assign to the monitoring and control system specialist radiological protection personnel, with instrumentation, with temporary work areas and the necessary operating and communication procedures to detect, separate and isolate any radioactive materials that may be detected.
- Provide its personnel with basic radiological protection and monitoring training appropriate to their activities and inform them of the characteristics of the company's monitoring and control system.

b) In connection with cross-border movements, imports or trade within the European Union involving metal materials:

- Require from the shipper an inspection certificate for the goods, issued by a reputable goods inspection and control body or agency, which states that his facilities have adequate radiation monitoring and control systems for the metal materials shipped and that the goods have been subjected to radiation monitoring.

- Not to unload on Spanish territory shipments that do not have the certificate referred to in the previous point.

c) Undertake, by itself or in collaboration with ENRESA, the necessary actions to return any detected radioactive materials to the foreign shipper.

d) Report immediately to the CSN the detection of radioactive material in a shipment of metal materials or in final products, using the format set out in annex 2.

e) Adopt the measures required to prevent the dispersion of the radioactive material.

f) Sign a contract with ENRESA for the handling of the radioactive materials in accordance with the provisions of point 6.3.a).

g) Transfer detected radioactive material to ENRESA.

h) Collaborate on radiological protection training and information campaigns among the employees of companies in the metal recovery and metal smelting sector.

6. Actions in the event of the detection of radioactive material.

6.1 The signatory company shall do the following:

a) In the case of detection of radioactive material in a consignment of metal materials that arrives at the facility.

- Immobilise the consignment in the facility in which it was detected.
- Notify the specialist radiological protection personnel who, using the appropriate radiation protection procedures, will:

  * Inspect the consignment in detail until they identify and separate the part or parts that contain the radioactive material.
  * Evaluate the nature and level of radioactivity that they contain.
  * Safely isolate the radioactive material.
  * Prepare a report describing the actions taken, their results and whether the radioactive material is exempt from nuclear regulation or must be transferred to ENRESA, in accordance with the criteria specified by the Transfer Authorisation.

- Inform the Nuclear Safety Council using the format laid out in annex 2 and forward the conclusions of the report prepared by the radiation protection specialists.
- Transfer the radioactive material to ENRESA as required by the Transfer Authorisation.
- Hold the radioactive material in safe conditions until it is removed by ENRESA.

b) Where radioactive material is detected in the process (pouring test in the case of a smelting plant), the signatory company shall do the following:

- Take samples of all the final products and perform an analysis on them.
- If the concentrations measured in these final product samples exceed the exemption levels laid down in appendix III of Directive 96/29/EURATOM, the person in charge of the facility shall *immediately*:
  ➢ Halt all contaminated phases of the process.
  ➢ Suspend the release from the facility of final products that have been in contact with the contaminated phases of the process.
  ➢ Call in a duly authorised Radiological Protection Technical Unit, which will determine the extent of the contamination in the process line and its immediate surroundings.
  ➢ Report the situation to the Nuclear Safety Council and to the recipients of final products who may have been in contact with the contaminated phases of the process.

6.2 Upon receipt of a radioactive material detection report, the Nuclear Safety Council shall do the following:

a) If radioactivity was detected in metal materials:

- Instruct the signatory company to transfer the radioactive material to ENRESA in accordance with the Transfer Authorisation.
- Advise ENRESA that the radioactive material will be transferred to it in accordance with the Transfer Authorisation.

b) If radioactivity was detected in final products, the Nuclear Safety Council shall:

- Inform the Ministry of Industry and Energy and recommend a course of action.
- Issue such instructions and recommendations as it considers necessary having regard to the information provided by the signatory company.
- Order such actions by the technical personnel and support services of the Nuclear Safety Council as it considers necessary.

6.3 Upon receipt of the Nuclear Safety Council report, ENRESA shall do the following:

a)   Remove the radioactive materials in accordance with the transfer authorisation, for which it will sign the appropriate contract with the signatory company.

b)   Hold the radioactive material under safe conditions until a decision is made on how it is to be definitively disposed of, which may involve:
   -   Return to the supplier if the latter is a foreigner,
   -   Transfer to another authorised agency
   -   Disposal as radioactive waste
   -   Any other legally authorised form of disposal

c)   Give the signatory company the necessary support in completing the required formalities for returning the radioactive material to the shipper, if the latter is a foreigner.

7. Special actions

When, in the judgement of the Nuclear Safety Council, the situation resulting from the contamination produced by the dispersion of radioactive material in a facility so requires, the Ministry of Industry and Energy, taking urgent action on the basis of the prior report of the Nuclear Safety Council, may require the adoption of such exceptional measures as it considers appropriate, in coordination, where necessary, with the other competent public agencies or affected companies.

8. Apportionment of the costs

The costs arising from the implementation of the Protocol shall be borne in accordance with the following criteria:

a)   The costs arising from the disposal of the radioactive materials detected either in metal materials or in final products shall be borne by the signatory company, without prejudice to the entitlement of the latter, where applicable, to recover them from the supplier or shipper.

b)   The provisions contained in the foregoing (point 8a) shall not apply to costs arising from the disposal of the radioactive sources that have been detected in metal materials proceeding from the national territory, which shall be for the account of ENRESA, pursuant to the Second Additional Provision of Act 14/1999 of 4 May, concerning Public Rates and Prices for services provided by the Nuclear Safety Council.

c)   Actions carried out by the Nuclear Safety Council arising from the agreements contained in this Protocol shall entitle the Agency to recover from the signatory company the cost of the performance thereof which shall be calculated in accordance with the provisions of article 31 of Act 14/1999 of 4 May, concerning Public Rates and Prices for services provided by the Nuclear Safety Council.

PART 3. **INFORMATION THAT MUST BE INCLUDED IN THE DECLARATION FOR THE REGISTRATION OF FACILITIES OF COMPANIES THAT ARE SIGNATORIES TO THE PROTOCOL ON THE RADIATION MONITORING OF METAL MATERIALS**

1.   Name of signatory company

2.   Description of the facility

    2.1   Location

    2.2   Basic characteristics of the facility

    2.3   Description of the processes carried out at the facility

    2.4   Plans of buildings, roads, entrances, etc.

    3.2   Approximate average annual production

3.   Description of the monitoring and control system

    3.1   Automatic instrumentation

    3.2   Mobile instrumentation

    3.3   Process monitoring instrumentation

    3.4   List of procedures

    3.5   Brief description of the isolation area

    3.6   Intervention by local plant personnel or by a Radiological Protection Technical Unit

4.   Person responsible for radiation monitoring in the facility

5.   Express declaration of acceptance of the Protocol for Collaboration on Radiation Monitoring of Metal Materials signed by the authorised signatory of the company.

PART 4.   REPORT ON THE DETECTION OF RADIOACTIVE MATERIAL TO THE NUCLEAR SAFETY COUNCIL

On the _____ day , at _____, at the facilities of the company

_____

_____

_____

the presence of radioactive material was detected in:

❑   a batch of scrap
❑   a batch of ingots
❑   a batch of semi-finished metal product
❑   the final products
❑   other items, to be specified
❑   photographic information is enclosed

coming from:

_____

supplied by:

_____

transported by:

_____

The batch was isolated in the designated isolation area and the radiological protection specialists
❑   of the facility
❑   of the radiological protection technical unit
have carried out a preliminary inspection of the batch and consider that it contains:

_____ items superficially contaminated with concentrations of
_____ Bq/cm^2 βγ emitters
_____ Bq/cm^2 α emitters
_____ sources housed   YES   NO in its shielding
Other items, to be specified_____

A contact dose rate of _____ μSv/h was measured and at a distance of 1 metre
_____ μSv/h was measured and the following protection measures were adopted:

❑   Identification of the storage area
❑   Isolation of the contaminated material
❑   Installation of additional shielding
❑   Decontamination of the contaminated material

_____ on the _____ of _____   of

For the signature
Company:

ANNEX 6

IAEA CODE OF CONDUCT ON THE SAFETY AND SECURITY OF RADIOACTIVE SOURCES
SEPTEMBER 2000

The IAEA's Member States

Noting that radiation sources are used throughout the world for a wide variety of beneficial purposes, e.g. in industry, medicine, research, agriculture and education,

Aware that their use involves risks due to radiation exposure,

Aware that these risks must be restricted and protected against through the application of appropriate radiation safety standards,

Aware that there have been a number of accidents with serious, even fatal, consequences during the use of radiation sources,

Recognising that such accidents may have an adverse impact on individuals and on the environment,

Recognising the importance of fostering a safety culture in all organisations and among all individuals engaged in the regulatory control or in the management of radiation sources,

Recognising the need for effective and continuous regulatory control, both within States and in situations involving the transfer of radiation sources between States,

Noting that serious accidents have occurred during the use of radiation sources, in particular radioactive sources, as a result of ineffective, or lapses in the continuity of, regulatory control, or as a result of lapses in management control during extended periods of storage,

Recognising that most of these accidents have been caused by the use of radioactive sources, including accidents involving orphan sources,

Recognising that a number of States may lack appropriate infrastructure for the safe management of radioactive sources, and that consequently exporting States should take due care in authorising exports,

Recognising the need for technical facilities, including appropriate equipment and qualified staff, to ensure the safe and secure management of radioactive sources,

Noting that the International Basic Safety Standards for Protection against Ionising Radiation and for the Safety of Radiation Sources contain recommendations for protection against exposure to ionising radiation and for the safety and security of radioactive sources,

Recalling the IAEA's Safety Requirements document on Legal and Governmental Infrastructure for Nuclear, Radiation, Radioactive Waste and Transport Safety,

Taking account of the provisions of the Convention on Early Notification of a Nuclear Accident (1986) and of the provisions of the Convention on Assistance in the Case of Nuclear Accident or Radiological Emergency (1986),

Taking account of the provisions of the Joint Convention on the Safety of Spent Fuel Management and on the Safety of Radioactive Waste Management (1997), in particular those provisions which relate to the transboundary movement of radioactive waste and to the possession, remanufacturing or disposal of disused sealed sources,

Recognising the global role of the IAEA in the areas of nuclear and radiation safety and the safety of radioactive waste management and disposal, and

Taking account of the "Categorisation of Radiation Sources" in the Annex to Attachment ... to IAEA document GOV/2000/..../GC(44)/...,

DECIDE that the following Code of Conduct should serve as guidance to States for - *inter alia* - the development and harmonisation of policies, laws and regulations on the safety and security of radioactive sources.

I.   SCOPE AND OBJECTIVE

1.   This Code applies to all radioactive sources that may pose a significant risk to health and the environment.   In implementing this Code, States should give highest priority to those radioactive sources which pose the most significant risks, i.e. the radioactive sources belonging to Category 1 of the IAEA's "Categorisation of Radiation Sources". However, in doing so, States should also devote appropriate attention to the regulation of radioactive sources other than those belonging to Category 1.

2.   This Code does not apply to the control of nuclear materials as defined in the Convention on the Physical Protection of Nuclear Materials.

3.   This Code also does not apply to radioactive sources within military or defence programmes. However, such sources should be managed in accordance with the principles of this Code.

4.   The objective of this Code is to achieve and maintain a high level of safety and security of radioactive sources through the development, harmonisation and enforcement of national policies, laws and regulations, and through the fostering of international co-operation.   In particular, this Code addresses the establishment of an adequate system of regulatory control from the production of radioactive sources to their final disposal, and a system for the restoration of such control if it has been lost.

5.   This Code relies on existing international standards relating to legal and governmental infrastructure for nuclear, radiation, waste and transport safety and to the control of radioactive sources. It is intended to complement existing international standards in these areas.

6.   In implementing this Code, States should emphasise and reinforce to manufacturers, suppliers, users and those managing disused sources their responsibilities for the safety and security of radioactive sources.

II. DEFINITIONS

7.   For the purposes of this Code:

"authorisation" means a permission granted in a document by a regulatory body to a legal person who has submitted an application to manufacture, supply, receive, store, use, transfer, import, export, transport, maintain or dispose of radioactive sources. The authorisation can take the form of a registration or a licence.

"disused source" means a radioactive source no longer intended to be used for its original purpose.

"management" means all activities, administrative and operational, that are involved in the manufacture, supply, receipt, storage, use, transfer, import, export, transport, maintenance or disposal of radioactive sources.

"orphan source" means a source which poses sufficient radiological hazard to warrant regulatory control but is not under regulatory control, either because it has never been under regulatory control, or because it has been abandoned, lost, misplaced, stolen or transferred without proper authorisation.

"radiation source" means a radiation generator, or a radioactive source or other radioactive material outside the nuclear fuel cycles of research and power reactors.

"radioactive source" means radioactive material that is permanently sealed in a capsule or closely bonded and in a solid form, excluding material within the nuclear fuel cycles of research and power reactors. It also includes any radioactive material released if the source is leaking or broken.

"regulatory body" means any body or bodies on which a State has conferred legal authority to regulate any aspect of the safety and security of radioactive sources, including legal authority to grant authorisations.

"regulatory control" means any form of control applied to facilities or activities by a regulatory body for reasons related to radiation protection or to the safety and security of radioactive sources.

"safety" means measures intended to minimise the likelihood of accidents with radiation sources and, should such an accident occur, to mitigate its consequences.

"security" means measures to prevent unauthorised access to, and loss, theft and unauthorised transfer of, radioactive sources.

III. BASIC PRINCIPLES

GENERAL

8. Every State should, in order to protect human health and the environment, take the appropriate steps necessary to ensure that the radioactive sources within its territory, or under its jurisdiction or control, are:

   (a) fit for purpose;
   (b) safely managed during their useful lives and at the end of their useful lives; and
   (c) not stored for extended periods of time in facilities not designed for the purpose of such storage.

9. Every State should establish an effective national legislative and regulatory system of control over the management of radioactive sources and over any other activity involving radioactive sources which entails a significant risk to individuals or the environment. Such a system should:

   (a) place the prime responsibility for the safe management of radioactive sources on the persons being granted the relevant authorisations;
   (b) minimise the likelihood of a loss of control;
   (c) provide for rapid response for the purpose of regaining control over sources that are no longer under control;
   (d) foster ongoing communication between the regulatory body and users; and
   (e) provide for its continual improvement.

10. Every State should ensure that appropriate facilities and services for radiation protection and safety are available to, and used by, the persons who are authorised to manage radioactive sources or undertake any other activity with radioactive sources within its territory. Such facilities and services should include those needed for:

(a) searching for missing sources and securing found sources;

(b) intervention in the event of an accident involving a radioactive source;

(c) personal dosimetry and environmental monitoring; and

(d) the calibration and intercomparison of radiation monitoring equipment.

11. Every State should ensure that adequate arrangements are in place for the appropriate training of the staff of its regulatory body, its customs officers, its police and the staff of other law enforcement agencies.

12. Every State should encourage bodies or persons likely to encounter orphan sources during the course of their operations to implement appropriate monitoring programmes to detect such sources.

LEGISLATION AND REGULATIONS

13. Every State should establish legislation and regulations that:
    (a) prescribe and assign governmental responsibilities for the safety and security of radioactive sources;
    (b) provide for the effective control of radioactive sources;
    (c) specify the requirements for protection against exposure to ionising radiation; and
    (d) specify the requirements for the safety and security of radioactive sources.

14. Such legislation and regulations should include, in particular:

    (a) the establishment of a regulatory body whose regulatory functions are effectively independent of other functions if that body is involved in both the management of radioactive sources and in their regulation. This body should have the powers listed in paragraphs 15 to 17;

    (b) measures, commensurate with the risks, to protect individuals and the environment from the deleterious effects of radiation;

    (c) administrative requirements relating to:
        (i) the authorisation of the management of radioactive sources; and
        (ii) the notification to the regulatory body, as appropriate, by an authorised person of actions involved in the management of such sources and of any other activity in relation to such sources which may engender a significant risk to individuals or the environment;

    (d) provisions for exemption, as appropriate, from these administrative requirements;

    (e) managerial requirements, in particular relating to the establishment of adequate policies, procedures and measures for the control of radioactive sources;

    (f) security measures to prevent, protect against, and ensure the timely detection of, the theft, loss or unauthorised use or removal of radioactive sources during all stages of management;

    (g) requirements relating to the verification of safety, through: safety assessments; monitoring and verification of compliance; and the maintenance of appropriate records; and

    (h) the imposition of appropriate penalties;

REGULATORY BODY

15. Every State should ensure that the regulatory body established by its legislation has the authority to:

   (a) establish regulations and issue guidance relating to the safety and security of radioactive sources;

   (b) require those who intend to use radioactive sources to seek an authorisation , and to submit a safety assessment when one is deemed necessary in the light of the risks posed;

   (c) obtain any relevant information from an applicant for an authorisation;

   (d) issue, amend, suspend or revoke, as necessary, authorisations for:
      (i) the management of radioactive sources; and
      (ii) any other activity involving such sources which may engender a risk to individuals or the environment;

   (e) attach clear and unambiguous conditions to the authorisations issued by it, including conditions relating to:
      (i) responsibilities;
      (ii) minimum operator competencies;
      (iii) minimum equipment performance criteria (including radioactive source requirements);
      (iv) requirements for emergency procedures and communication links;
      (v) work procedures to be followed;
      (vi) maintenance of equipment and sources; and
      (vii) the adequate management of disused sources, including, where applicable, agreements regarding the possible return of decayed/disused sources to a supplier;

   (f) obtain any relevant and necessary information from the holder of an authorisation;

   (g) enter premises of authorised users to undertake inspections, according to established procedures, to verify compliance with regulatory requirements;

   (h) enforce regulatory requirements;

   (i) monitor, or request other authorised bodies to monitor, at appropriate checkpoints for the purpose of detecting orphan sources;

   (j) ensure that corrective actions are taken when a radioactive source is in an unsafe condition;

   (k) provide, on a case-by-case basis, to the holder of an authorisation and the public any information that is deemed necessary in order to protect individuals and the environment;

   (l) liaise and co-ordinate with other governmental bodies and relevant non-governmental bodies within the State, and also with international bodies and regulatory bodies in other States, in order to seek guidance, information and assistance relevant to the safe and secure management of radioactive sources; and

   (m) establish criteria for intervention in emergency situations.

16. Every State should ensure that its regulatory body:

   (a) is staffed by qualified personnel; and

   (b) has the financial resources and the facilities and equipment necessary to undertake its functions in an effective manner.

17. Every State should ensure that its regulatory body:

   (a) establishes procedures for dealing with applications for authorisation;

   (b) ensures that, before the receipt of a radioactive source is authorised:
      (i) arrangements have been made for its safe management once it has become a disused source; and
      (ii) financial provision has been made for its safe management once it has become a disused source.

(c) maintains appropriate records of holders of authorisations in respect of radioactive sources, with a clear indication of the type(s) of the radioactive sources that they are authorised to use, and appropriate records of the transfer and disposal of the radioactive sources on termination of the authorisation;

(d) establishes systems for ensuring that, where practicable, both radioactive sources belonging to Categories 1 and 2 of the IAEA's "Categorisation of Radiation Sources", and their containment, are marked with an appropriate sign to warn members of the public of the radiation hazard, but where this is not practicable, at least the containment is so marked.

(e) establishes systems for ensuring that, where practicable, radioactive sources belonging to Categories 1 and 2 of the IAEA's "Categorisation of Radiation Sources" are identifiable and traceable;

(f) ensures that inventory controls are conducted on a regular basis by the holders of authorisations;

(g) carries out both announced and unannounced inspections at a frequency determined by past performance and the risks presented by the radioactive source;

(h) takes enforcement actions, as appropriate, to ensure compliance with regulatory requirements;

(i) ensures that the regulatory principles and criteria remain adequate and valid and take into account, as applicable, operating experience and internationally endorsed standards and recommendations;

(j) requires the prompt reporting by authorised persons of loss of control over, and of incidents in connection with, radioactive sources;

(k) prescribes appropriate levels of training for manufacturers, suppliers and users of radioactive sources;

(l) requires authorised persons to prepare appropriate emergency plans;

(m) is prepared, or has established provisions, to recover orphan sources and to deal with radiological emergencies and has established appropriate response plans and measures;

(n) is prepared, in respect of any radioactive source whose export it has authorised, to provide, upon request, information relating to its safe management.

IMPORT AND EXPORT OF RADIOACTIVE SOURCES

18. Every State intending to import a radioactive source belonging to Categories 1 and 2 of the IAEA's "Categorisation of Radiation Sources" should consent to its import only if the State has the technical and administrative capability needed to manage the source in a manner consistent with the provisions of this Code.

19. A State should allow for re-entry into its territory of disused radioactive sources if, in the framework of its national law, it has accepted that they be returned to a manufacturer qualified to receive and possess the disused radioactive sources.

20. Any State which authorises the export of a radioactive source should take appropriate steps to ensure that such export is undertaken in a manner consistent with existing international standards relating to the safe transport of radioactive materials.

ROLE OF THE IAEA

21. The IAEA should:

(a) continue to collect and disseminate information on laws, regulations and technical standards relating to the safe and secure management of radioactive sources, develop and establish relevant technical standards and provide for the application of these standards at the request of any State, inter alia by advising and assisting on all aspects of the safe and secure management of radioactive sources; and

(b) in particular, implement the measures approved by its governin[g]
pursuant to its Action Plan on the Safety of Radiation Sources a[nd]
Radioactive Materials.

DISSEMINATION OF THE CODE

22. Every State should inform public and private organisations and persons inv[olved in]
management of radioactive sources, as appropriate, of the measures it has [taken to]
implement this Code and should take steps to disseminate that information widely.

Printed at United Nations, Geneva
GE.02-31307–June 2002–2,110

ECE/TRADE/278

United Nations publication
Sales No. E.01.II.E.22

ISBN 92-1-116789-2